Black Silicon

Technology, opportunities and challenges

Online at: https://doi.org/10.1088/978-0-7503-3247-7

Black Silicon

Technology, opportunities and challenges

N M Ravindra

Department of Physics, New Jersey Institute of Technology, Newark, NJ, United States of America

Priyanka Singh

Advanced Materials and Devices Metrology Division, CSIR-National Physical Laboratory, New Delhi, India

IOP Publishing, Bristol, UK

ISBN 978-0-7503-3247-7 (ebook)
ISBN 978-0-7503-3245-3 (print)
ISBN 978-0-7503-3248-4 (myPrint)
ISBN 978-0-7503-3246-0 (mobi)

DOI 10.1088/978-0-7503-3247-7

Version: 20250701

IOP ebooks

British Library Cataloguing-in-Publication Data: A catalogue record for this book is available from the British Library.

Published by IOP Publishing, wholly owned by The Institute of Physics, London

IOP Publishing, No.2 The Distillery, Glassfields, Avon Street, Bristol, BS2 0GR, UK

US Office: IOP Publishing, Inc., 190 North Independence Mall West, Suite 601, Philadelphia, PA 19106, USA

To the late Professor Abdus Salam and the Abdus Salam Center for Theoretical Physics

Contents

Preface

The year 2024 marked the 200th anniversary of the discovery of amorphous silicon by Jöns Jacob Berzelius, a Swedish chemist [1–3]. The amorphous silicon was formed by reacting potassium fluorosilicate with potassium and purifying the product through repeated washing [4, 5]. The galena (lead sulfide) point-contact rectifier was the first semiconductor device invented by Ferdinand Braun in 1874 [6], followed by the radio wave detector, invented by Jagadish Chandra Bose in 1901 [7]. The first application of silicon in microelectronics, in the form of a silicon transistor, was reported in 1954 by Morris Tanenbaum [8], 130 years after the discovery of silicon. It was followed by the production of the commercial silicon transistor by Gordon Teal [9] at Texas Instruments. The invention of the bipolar integrated circuit by Jack Kilby [10, 11] and Robert Noyce [11, 12] led to the birth of silicon microelectronics. The subsequent development of complementary metal-oxide semiconductor (CMOS) technology and the birth of modern microelectronics was catalyzed by the invention of the metal-oxide semiconductor field-effect-transistor (MOSFET) by Mohamed Atalla and Dawon Kahng in 1959 [13, 14] at Bell Telephone Laboratories (Bell Labs). In the early 1960s, silicon became the standard material of choice for microelectronics.

Today the silicon microelectronics industry is firmly established as a critical infrastructure to address the demands of the various applications in the advanced global economy, for example:

- *Advancements in technology*: Advanced component, system, and packaging architectures such as in chiplets, chip-on-wafer-on-substrate (CoWoS), lab-on-a-chip (LOC), system-on-a-chip (SOC), XYZ on a chip, micro-electro-mechanical structures (MEMS), sensor architecture and integration, and the associated thermal management schemes.
- *Artificial intelligence (AI) and machine learning (ML):* The exponential increase in adoption and usage of high-performance computing (HPC) and efficient semiconductors, while addressing thermal management.
- *Automotive industry*: Increased use of electric vehicles, autonomous driving, advanced driver-assistance systems (ADAS) that require significant use of advanced semiconductors, and energy storage, for example as anodes in batteries.
- *Digital transformation*: The increased demand for semiconductor solutions in several sectors of the global economy such as healthcare, manufacturing, retail electronics, and transportation.
- *Energy*: Silicon is firmly established as the most dominant semiconductor in the manufacture of solar cells with a market share of ~95%; energy storage and thermoelectrics are other areas of its applications.
- *High performance computing*: Solutions to address the exponential growth in data centers, cloud computing, advanced semiconductor solutions in graphics processing units (GPUs), high-bandwidth memory, and solid-state drives (SSDs) and the associated thermal management solutions.

- *5G (fifth generation of cellular technology) and connectivity*: Integration of microelectronics and photonics that can enable rapid, highly efficient, and compact systems with reduced demands on power while operating at extremely high speeds/frequencies.
- *Sensors*: Silicon is widely used in the manufacture of a variety of sensors including biosensors, CMOS image sensors, force/pressure sensors, gyroscopes, motion sensors, particle radiation sensors, position sensors, temperature sensors, and x-ray detectors.

Industry players such as Nvidia, AMD, Axelera AI, Blaize, Broadcom, Horizon Robotics, Intel, and Qualcomm continue to seek solutions that address the needs of integration of microelectronics, photonics, and thermal management, as well as the various semiconductors that are utilized in the manufacturing of these chips. For example, packaging of these complex chips requires solutions to counter thermally induced stress at the systems level while also addressing differences in thermal expansion coefficients at the component level.

The relative abundance of silicon in nature, coupled with the maturity and reliability of its technology, has been a huge asset to silicon microelectronics and solar photovoltaics. While there are commonalities in the problems and issues that are faced by the silicon microelectronics and the silicon solar cell industry, the principal emphasis in the silicon solar cell industry is the dollar per watt of installed solar panels. Unlike silicon photovoltaics, the indirect bandgap of silicon and the wafer cost are less of an issue in silicon microelectronics.

The renewable energy industry, particularly photovoltaics, is under significant pressure to maximize the return on investments (ROIs), while countering the geopolitical problems that are associated with global supply-chain problems, tariffs etc. Silicon accounts for ~95% of the semiconductor materials used in the manufacture of solar cells in the world today. Silicon exists in several forms, including the following: single crystal, multi-crystalline, polycrystalline, microcrystalline [15], and amorphous phase [16]. With its indirect bandgap, single crystal silicon (energy gap, $E_G = 1.17$ eV) exhibits lower photon absorption coefficients compared to direct bandgap semiconductors such as copper indium gallium diselinide (CIGS), E_G = 1.0–1.7 eV, cadmium telluride (CdTe), $E_G = 1.45$ eV, copper zinc tin sulfide (CZTS), E_G = 1.4–1.6 eV, gallium arsenide (GaAs), $E_G = 1.42$ eV, and hydrogenated amorphous silicon (a-Si:H), E_G = 1.5–2.0 eV [17–19]. Nanocrystalline silicon is also referred to as microcrystalline silicon (μc-Si) and is a form of porous silicon. Silicon in two dimensions, also known as silicene, has been reported to have a quasi-direct bandgap of ~1.11 eV and is anticipated to have applications in optoelectronics [20].

One of the several approaches to reduce the cost of manufacturing silicon solar cells relates to the reduction in the thickness of the substrates. The cost of a crystalline silicon wafer is estimated to be ~10%–20% of the overall cost of a solar cell. Factors such as the country of origin, associated tariffs and taxes, dimensions of the wafers (size, quality, thickness), architecture of the cell, manufacturing process, as well as the efficiency and power output of the cell contribute to the cost of the cell. Today's silicon solar cells utilize silicon wafers that are 160–200 μm thick and have a dimension of 18.2 cm × 18.2 cm (M10 Wafers) or G12R rectangular wafer format (18.2 cm x 21.0 cm). Due to the increased absorption coefficient, amorphous silicon solar cells can have an active layer that is 1 μm thick.

The quest for silicon photonics [21–23] has been around for some time. The repeatability and reproducibility of a silicon photonic device will be the driver for the integration of silicon microelectronics-photonics technology and subsequent applications in optical communications. Intel has recently demonstrated the first fully integrated optical I/O chiplet [24] that will facilitate high-speed data transmission. The enhancement in the absorption coefficient in black silicon (BSi) [25–31] has been demonstrated in numerous studies in the literature. SiOnyx [32] has pioneered the design, development, and manufacture of high-performance image sensors utilizing its proprietary BSi technology. Researchers from Aalto University in Finland have demonstrated high efficiency BSi solar cells [33]. Research on BSi continues to progress as it finds applications in numerous areas including energy, healthcare, thermal management etc. From a manufacturing perspective, BSi has significant potential and it needs to be exploited. The advantages of BSi are numerous including the fact that its integration with crystalline silicon is facile, offering thereby an all-silicon solution. In photovoltaics, BSi plays a dual role as an antireflection coating as well as the substrate for device manufacture.

Newark, New Jersey, 11 May, 2025

References

[1] Jöns Jacob Berzelius *Wikipedia* https://en.wikipedia.org/wiki/Jöns_Jacob_Berzelius (Accessed: 10 May 2025)

[2] The Element Silicon *JLab* https://education.jlab.org/itselemental/ele014.html (Accessed: 10 May 2025)

[3] Stewart D n.d. Silicon element facts *Chemicool* www.chemicool.com/elements/silicon.html (Accessed: 10 May 2025)

[4] Weeks M E 2003 *Discovery of the Elements* (Whitefish, MT: Kessinger) p 162

[5] Emsley J 2002 *Nature's Building Blocks: An A–Z Guide to the Elements* (Oxford: Oxford University Press) p 387

[6] 1874: Semiconductor point-contact rectifier effect is discovered *Computer History Museum* www.computerhistory.org/siliconengine/semiconductor-point-contact-rectifier-effect-is-discovered/ (Accessed: 10 May 2025)

[7] MacNeil J 2019 JC Bose patents radio wave detector, September 30, 1901 *EDN* https://www.edn.com/jc-bose-patents-radio-wave-detector-september-30-1901/ (Accessed: 10 May 2025)

[8] Silicon transistor *ETHW* https://ethw.org/Silicon_Transistor (Accessed: 10 May 2025)

[9] Teal G 1955 Method of producing a semiconductor element *US Patent* No. 2703296A

[10] History of semiconductors *Hitachi High-Tech* https://hitachi-hightech.com/global/en/knowledge/semiconductor/room/about/history.html (Accessed: 10 May 2025)

[11] 1959: Practical monolithic integrated circuit concept patented *Computer History Museum* https://www.computerhistory.org/siliconengine/practical-monolithic-integrated-circuit-concept-patented/ (Accessed: 10 May 2025)

[12] Noyce R N 1961 Semiconductor device-and-lead structure *US Patent* 2981877

[13] Leibson S 2025 A brief history of the MOS transistor. Part 1: Early Visionaries *Electronic Engineering Journal* https://www.eejournal.com/article/a-brief-history-of-the-mos-transistor-part-1-early-visionaries/ (Accessed: 11 May 2025)
[14] Bassett R K 2025 *To the Digital Age: Research Labs, Start-Up Companies, and the Rise of MOS Technology* Johns Hopkins Studies in the History of Technology (Baltimore, MD: Johns Hopkins University Press)
[15] Basore P A 1994 Defining terms for crystalline silicon solar cells *Prog. Photovoltaics Res. Appl.* **2** 177–9
[16] Kang H 2021 Crystalline silicon vs amorphous silicon: the significance of structural differences in photovoltaic applications *IOP Conf. Series: Earth Environ. Sci.* **726** 012001
[17] Ravindra N M and Demichelis F 1985 Cody disorder: absorption–edge relationships in hydrogenated amorphous silicon *Phys. Rev.* B*32 6591–5*
[18] Demichelis F 1986 Optical properties of hydrogenated amorphous silicon *J. Appl. Phys.* **59** 611–8
[19] Ravindra N M *et al* 1986 Low-temperature optical properties of hydrogenated amorphous silicon *Mater. Lett.* **4** 343–9
[20] Borlido P *et al* 2018 The ground state of two-dimensional silicon *2D Mater.* **5** 035010
[21] Reynolds M 2025 Controlling light: is silicon photonics an emerging front in US–China tech competition? *CSIS* https://csis.org/analysis/controlling-light-silicon-photonics-emerging-front-us-china-tech-competition (Accessed: 11 May 2025)
[22] Gupta R *et al* 2025 The integration of microelectronic and photonic circuits on a single silicon chip for high-speed and low-power optoelectronic technology *Nano Mater. Sci.* https://doi.org/10.1016/j.nanoms.2024.04.011
[23] Sekhar S *et al* 2024 Roadmapping the next generation of silicon photonics *Nat. Commun.* **15** 751
[24] Intel 2024 Intel demonstrates first fully integrated optical I/O chiplet *Intel Newsroom* https://newsroom.intel.com/artificial-intelligence/intel-unveils-first-integrated-optical-io-chiplet (Accessed: 11 May 2025)
[25] Jansen H *et al* 1995 The black silicon method: a universal method for determining the parameter setting of a fluorine-based reactive ion etcher in deep silicon trench etching with profile control *J. Micromechan. Microeng.* **5** 115
[26] Ravindra N M, Marthi S R and Sekhri S 2016 *Black Silicon: Processing, Properties and Applications* (New York: Momentum)
[27] Marthi S R 2017 Black silicon—optical properties and applications *PhD Dissertation* New Jersey Institute of Technology, Newark, NJ
[28] Sekhri S 2015 Optical properties of black silicon—an analysis *MS Thesis* (Newark, NJ: New Jersey Institute of Technology)
[29] Marthi S R and Ravindra N M 2016 Optical properties of passivation layers on black silicon *Proc. Materials Science and Technology Conf. and Exhibition (Salt Lake City)* pp 1141–52
[30] Marthi S R, Sekhri S and Ravindra N M 2015 Optical properties of black silicon: an analysis *JOM* **67** 2154–9
[31] Sarnet T *et al* 2013 Black silicon for photovoltaic cells: towards a high-efficiency silicon solar cell *EU PVSEC 2013, 28th European Photovoltaic Solar Energy Conf. and Exhibition*
[32] *SIONYX* https://sionyx.com/ (Accessed: 11 May 2025)
[33] Savin H *et al* 2015 Black silicon solar cells with interdigitated back-contacts achieve 22.1% efficiency *Nat. Nanotechnol.* **10** 624–8

Acknowledgements

As I, N M Ravinda, cherish my privileged intellectual past, during the years 1982–85, I recollect my extensive collaborations in research on amorphous silicon with the research groups at Laboratoire de Spectroscopie II, Université de Montpellier (Professor Madame S Robin, J P Ferraton, C Ance, and the team in Montpellier), Université Pierre et Marie Curie (Professor Minko Balkanski, J F Mhoronge, M Jouanne, M A Kanehisa, and the team in Paris), Politecnico di Torino (the late Professor Francesca Demichelis, Elena Tresso, Alberto Tagliaferro, and the team in Torino), and Elettrorava (Paulo Rava). The award of visiting fellowships at the International Center for Theoretical Physics in Trieste (heartfelt thanks to the late Professor Abdus Salam, the late Professor G Furlan, and the team) is acknowledged with profound gratitude. My discussions with the late David Adler (MIT) and the late Arun Madan (MV Systems) are eternally engraved in my thought process.

Looking forward, 29 January 2026 will mark the 100th birthday of Professor Abdus Salam, and 19 July 2026 will be the tenth anniversary since Professor Furlan passed away. It has been 40 years since I left Trieste. As I recall my days in Trieste, I could not imagine a better place to practice science—indeed, Trieste is the Mecca to preach, practice, and share science—without borders.

P Singh thanks the New Jersey Institute of Technology and the National Physical Laboratory. She thanks her Gurus for their enduring guidance and wisdom, as well as her family for their constant love and support.

The authors thank Ms Judith Sheft, Executive Director of the New Jersey Commission of Science, Innovation and Technology, and Professors Michael Ehrlich and Cesar Bandera of the NJIT School of Management for their support of the Black Silicon Project at NJIT through grants from NSF I-Corps. The completion of this book overlaps with the authors' participation in the National NSF I-Corps™ Desert and Pacific Winter Cohort 2025 and its completion. The authors thank the National Science Foundation for its support.

The authors thank Ms Mia Foulkes and the team at IOP for the constant help, guidance, and support.

Author biographies

Dr Nuggehalli M Ravindra

Dr Nuggehalli M Ravindra (Ravi) is a Professor of Physics at the New Jersey Institute of Technology (NJIT). Previously, Ravi was the Chair of the Department of Physics and Director of the Interdisciplinary Program in Materials Science and Engineering at NJIT. A co-author of more than 350 papers and eight books, Ravi's research interests are in the areas of education, electronic materials, devices and structures, energy storage, renewable energy, and semiconductors.

Dr Priyanka Singh

Dr Priyanka Singh is a Woman Scientist with the National Physical Laboratory, New Delhi, India. Previously, she worked as a Research Scientist with Professor N M Ravindra, in the Department of Physics at New Jersey Institute of Technology (NJIT). Priyanka is a co-author of more than 13 papers and one book in the field of solar cells and semiconductors.

IOP Publishing

Chapter 1

Silicon—evolution

A brief overview of the physical, chemical, electronic, thermal, and mechanical properties of silicon is presented in this chapter. Some of the trends in the scaling of silicon-related transistors and the recent initiatives by the Defense Advanced Research Project Agency (DARPA) on the 3D Heterogeneous Integration (3DHI) Next-Generation Microelectronics Manufacturing Program (NGMM) are highlighted.

1.1 Introduction

Much of the evolution in semiconductor science and technology can be attributed to the abundance, growth, maturity, reliability, stability, and applications of silicon (Si). It is only natural that Si continues to be a major player, with over 90% of photovoltaic modules made using Si. The prominence of Si has grown over the decades due to its unique properties, abundance, and adaptability in various applications, in addition to its cost-effectiveness. The semiconductor market is expected to be valued at $1 trillion by 2030 [1]. Estimates suggest that silicon will continue to hold a significant market share of ~70%–80% of the total semiconductor market in 2030.

Silicon was first discovered by the Swedish chemist J J Berzelius in 1824. Silicon is the second most abundant element in the Earth's crust (about 28% by weight), after oxygen. It primarily occurs in the form of silicon dioxide (SiO_2) in rocks, sand, and quartz [2]. Silicon is classified based on its purity level, which directly influences its applications. Higher purity silicon is essential for electronic and photovoltaic applications, while lower purity grades are sufficient for metallurgical applications. Metallurgical grade silicon (MG-Si, purity ~98%–99%) is produced by the reduction of silicon dioxide in an electric arc furnace and is commonly used as an alloying agent in aluminum and steel production. Solar-grade silicon (SoG-Si, purity ~99.9999% or above) is obtained by purifying MG-Si using the Siemens process. Electronic-grade silicon (EG-Si, purity ~99.999 9999% or higher) is further purified

doi:10.1088/978-0-7503-3247-7ch1

from SoG-Si using processes such as zone refining or chemical vapor deposition (CVD). Monocrystalline silicon (c-Si) wafers are made using the Czochralski (CZ) or float-zone (FZ) methods. Monocrystalline silicon, a continuous, unbroken crystal structure, is used in high-performance applications such as computer microprocessors and high-efficiency solar cells. Unlike c-Si, amorphous silicon (a-Si) has no regular arrangement of atoms and is used in applications including thin-film solar cells (less efficient than c-Si) and some display technologies. The development of silicon-based integrated circuits (ICs) has historically allowed for the miniaturization of electronic devices, giving rise to the modern day computing era. Texas Instruments (TI) developed the first practical silicon transistor in 1954, revolutionizing semiconductor electronics. Silicon became the foundation of the microelectronics revolution, leading to the creation of microprocessors, which became the brains of computers and other electronic devices. The region in the southern part of California's San Francisco Bay area became the epicenter of semiconductor innovation and the birthplace of many of the world's leading companies, earning the nickname 'Silicon Valley'. In 1954, Bell Labs developed the first practical silicon solar cell, achieving a 6% efficiency in converting sunlight to electricity. Over the following decades, efficiencies improved through innovative technologies [2]. Modern c-Si based solar cells dominate the market with efficiencies of commercial silicon solar cells exceeding 22%. Research is ongoing to push the efficiency limits while integrating it with other materials (e.g. tandem cells with perovskites).

The potential role of Si in quantum computing is also being explored, with researchers working on silicon-based qubits [3], which could provide a pathway to scalable quantum processors. Advancements in stacking silicon layers vertically (3D architecture) aim to continue the trend of increasing transistor density (i.e. the number of transistors per unit area) despite the physical limitations of shrinking chip sizes.

1.2 Physical properties

Silicon is a metalloid with a unique set of physical properties that make it invaluable for applications in electronics, solar cells, and various industries [4]. In its pure form, it is a hard, brittle, grayish-black, crystalline solid with a metallic luster (figure 1.1

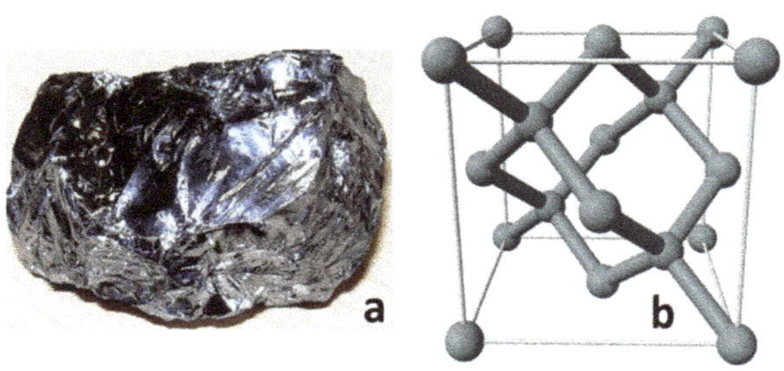

Figure 1.1. (a) Silicon as a crystalline solid. (b) Crystal structure of silicon. ((a) Reproduced from [5]. (b) Reproduced from [6]. These images have been obtained by the author(s) from the Wikimedia website, where they are stated to have been released into the public domain. They are included within this article on that basis.)

Table 1.1. Physical properties of silicon [5, 7, 8].

Atomic number	14
Atomic weight	28.085 u
Crystal structure	Diamond cubic
Melting point	1414 °C (2577 °F)
Boiling point	3265 °C (5909 °F)
Band gap	1.12 eV (at 300 K)
Refractive index	3.42 (at 589 nm wavelength)
Isotopes	Si-28 (92.23%), Si-29 (4.67%), Si-30 (3.1%)
Color	Metallic gray (crystalline form)

Table 1.2. Chemical properties of silicon [5].

Chemical symbol	Si
Electronic configuration	[Ne] $3s^2$ $3p^2$
Valence electrons	4
Oxidation states	+4, +2, −4
Electronegativity	1.90 (Pauling scale)
Ionization energy	First: 786.5 kJ mol^{-1}
Ionic radius	Si^{4+}: 40 pm
Common compounds	Silicon dioxide (SiO_2), silicon carbide (SiC), silicates
Standard electrode potential	−0.87 V (Si^{4+}/Si)

(a)). Silicon has a diamond cubic crystal structure, which contributes to its semiconducting properties. In this structure, each silicon atom is covalently bonded to four other silicon atoms in a tetrahedral arrangement, as shown in figure 1.1(b).

Silicon is relatively lightweight compared to many metals, which makes it suitable for large-scale industrial use and manufacturing, especially in thin, lightweight electronic devices. Table 1.1 highlights the key physical properties of silicon that make it suitable for use in semiconductors, electronics, and photovoltaics.

1.3 Chemical properties

Silicon is chemically inert at room temperature but reacts with halogens such as chlorine and fluorine to form silicon halides ($SiCl_4$) and dilute alkalis at higher temperatures. It forms SiO_2 when exposed to oxygen or air at elevated temperatures, which protects it from further oxidation. This property is critical in its use in semiconductor devices, in which the oxide layer acts as an insulator. It has four valence electrons, making it tetravalent and capable of forming four covalent bonds. Silicon is resistant to most acids except hydrofluoric acid (HF), which dissolves it to form hexafluorosilicic acid (H_2SiF_6). Table 1.2 summarizes the chemical properties of silicon.

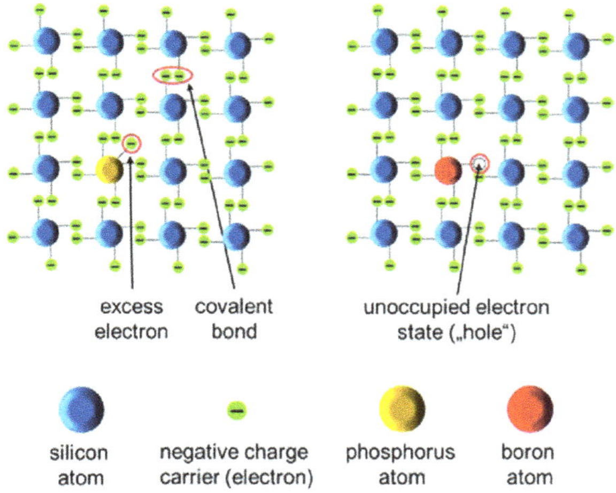

Figure 1.2. n-doped silicon with phosphorus (left) and p-doped with boron (right). (Reproduced with permission from [9]. Copyright Springer Nature 2020.)

1.4 Electronic properties

The electronic properties of Si are the foundation of modern electronics and semiconductor devices. Si is a semiconductor, having electrical conductivity between a conductor and an insulator. In its pure (intrinsic) form, it has relatively low conductivity at room temperature. However, the conductivity of Si can be increased by introducing impurities (doping) and can be made either n-type (with excess electrons as charge carriers) or p-type (with holes as charge carriers) as shown in figure 1.2. This tunable electrical conductivity is crucial for the manufacture of semiconductor devices such as diodes and transistors. In silicon, the mobility of electrons is higher than that of holes; therefore, electrons move more easily through the lattice compared to holes. When energy (such as heat or light) is applied to silicon, electrons in the valence band are excited to the conduction band, creating electron–hole pairs, which are essential for the modification of the electrical conductivity of silicon and its applications in semiconductor devices. In intrinsic silicon, electrons can recombine with holes, effectively reducing its electrical conductivity. The recombination rate is crucial in determining the efficiency of devices such as light emitting diodes (LEDs) and solar cells. Table 1.3 highlights the most important electronic properties of silicon.

1.5 Thermal properties

In electronic devices such as diodes, transistors, and ICs, the thermal properties of silicon are critical for heat dissipation and device reliability. Efficient thermal management extends the lifespan of electronic components, circuits, and systems. The significant thermal characteristics of Si are listed in table 1.4. The relatively high thermal conductivity of Si (149 W m^{-1} K^{-1}) transfers heat efficiently. Electronic devices such as power transistors and solar panels remain stable at high

Table 1.3. Electronic properties of silicon [4, 5].

Intrinsic carrier concentration	$\sim 1.5 \times 10^{10}$ cm^{-3} (at 300 K)
Electron mobility	1350 cm^2 V^{-1} s^{-1} (at 300 K)
Hole mobility	480 cm^2 V^{-1} s^{-1} (at 300 K)
Electrical conductivity	Varies with doping
Dielectric constant (permittivity)	11.7 (relative to vacuum)
Resistivity (Ω·cm)	Intrinsic silicon $\sim 2.3 \times 10^5$ Lightly doped (n or p) 10^0–10^2 Heavily doped (n or p) 10^{-3}–10^0
Recombination time	1–1000 μs (varies with purity and doping levels)
Doping types	n-type (donors—phosphorus, arsenic), p-type (acceptors—boron, gallium)
Effective mass of electrons m^*_e	$\approx 0.26 m_0$ m_0—free electron mass ($\sim 9.11 \times 10^{-31}$ kg)
Effective mass of holes (m^*_h)	$\approx 0.39 m_0$
Breakdown electric field	~ 300 kV cm^{-1}

Table 1.4. Thermal properties of silicon [5, 10].

Thermal conductivity	149 W m^{-1} K^{-1} (at 300 K)
Specific heat capacity	0.71 J g^{-1} K^{-1} (at 300 K)
Thermal expansion coefficient	2.6×10^{-6} K^{-1} (at 300 K)
Thermal diffusivity	88 mm^2 s^{-1} (at 300 K)
Debye temperature	645 K
Heat of fusion	50.55 kJ mol^{-1}
Heat of vaporization	384 kJ mol^{-1}
Specific energy of formation	−911 kJ mol^{-1} (SiO_2 formation)

temperatures due to the high melting point (1414 °C) of Si. Silicon has a specific heat capacity of 0.71 J g^{-1} K^{-1}; so, it requires a moderate amount of energy to raise its temperature. This is important for managing temperature changes in electronic devices, ensuring that they can operate safely over a wide range of temperatures. Furthermore, the low thermal expansion coefficient (2.6×10^{-6} K^{-1}) of Si indicates that it expands very little with changes in temperature. This minimizes the risk of thermal stress in devices that undergo temperature fluctuations, contributing to their durability and reliability in semiconductor applications. In addition, Si has high thermal diffusivity (88 mm^2 s^{-1}), which allows it to control heat more effectively, dissipating energy more uniformly across its surface. The Debye temperature for

Table 1.5. Mechanical properties of silicon [11].

Density	2.33 g cm^{-3} (at 300 K)
Hardness	7.0 Mohs
Young's modulus	130–185 GPa (depending on crystallographic orientation)
Poisson's ratio	0.22–0.28
Fracture toughness	0.9 MPa·$m^{1/2}$ (low, indicating brittleness)
Tensile strength	170 MPa (high strength in single-crystal form)
Compressive strength	1–2 GPa (depending on crystal orientation)
Shear modulus	65 GPa
Bulk modulus	98 GPa
Brittleness	High (breaks easily under mechanical stress)

heat capacity and phonon vibrations of silicon is 645 K, which is consistent in managing high thermal loads.

1.6 Mechanical properties

Silicon has numerous critical mechanical properties that make it suitable for use in a variety of high-tech applications, particularly semiconductors, electronics, and micro-electromechanical systems (MEMS). The key mechanical properties of Si are listed in table 1.5. The diamond cubic crystal structure of silicon significantly influences its mechanical properties. The Young's modulus (130–188 GPa) of Si depends on the crystallographic direction and makes it quite stiff and resistant to deformation. Its Poisson's ratio is around 0.22–0.28 which measures the tendency of Si to expand in directions that are perpendicular to the direction of compression. Furthermore, silicon is brittle and has a low fracture toughness, typically between 0.9 and 1 MPa·$m^{1/2}$. This makes it prone to cracking under mechanical loads. The tensile strength of silicon is direction-dependent (anisotropic) and typically ranges from 110 to 170 MPa, making it relatively strong for its weight but brittle. In addition, Si is a hard material with a Vickers hardness of about 850–1200 HV, which is comparable to some metals. It has a density of about 2.33 g cm^{-3}, which makes it much lighter than many metals, though denser than polymers. The shear modulus of silicon is around 50–70 GPa, indicating its resistance to change in shape under shear stress. Due to its brittleness and mechanical strength, silicon is widely used in MEMS and other applications [11]. However, it is also prone to fracture, limiting its mechanical uses under high stress or in structural applications. These properties limit the practical and usable dimensions, including thickness, of silicon wafers in manufacturing.

1.7 Scaling—present trends

Scaling of silicon refers to the process of shrinking the size of transistors on silicon chips, which allows for more transistors to fit on a chip, improving the performance and power efficiency, and reducing cost per transistor [12]. This has driven the progress in semiconductor technology over the past several decades, traditionally following Moore's law, which predicted that the number of transistors on a chip would double approximately every two years. However, as transistors approach atomic scales, the silicon scaling process faces significant challenges and has led to new strategies in semiconductor development. At extremely small scales, quantum mechanical phenomena such as electron tunneling become significant, leading to leakage current and increased power consumption. Smaller transistors and higher densities (number of transistors that can be placed on an area of a microchip) generate more heat, which is challenging to manage and can degrade performance or even cause damage to chips. Traditional photolithography techniques that are used to etch circuits onto silicon wafers face physical limitations when dealing with extremely small feature sizes. Extreme ultraviolet (EUV) lithography is being developed to address these challenges, but it is expensive and complex [13].

Silicon transistors have been scaled down to nodes as small as 3 nm, with further reductions (e.g. 2 nm nodes) on the horizon. However, this progression is significantly slower and more expensive than in previous decades. Leading semiconductor companies including TSMC, Samsung, and Intel are at the forefront of pushing silicon scaling further, although they are now exploring alternative strategies alongside traditional scaling. Fin field-effect transistors (FinFETs) have been the standard transistor architecture for nodes smaller than 22 nm. FinFETs are 3D transistors that provide better control over the channel and reduce leakage current. Gate-all-around (GAA) transistors are the next step beyond FinFETs and are expected to dominate at 3 nm and below. GAA transistors use stacked nanowires or nanosheets that further improve electrostatic control and reduce power consumption [14].

In recent years, DARPA has been working on the Next-Generation US Microelectronics Manufacturing (NGMM) [15] infrastructure. The goal of this program is to be able to create an infrastructure in support of prototyping the chips of tomorrow. This is illustrated in figure 1.3. Such efforts are anticipated to lead to

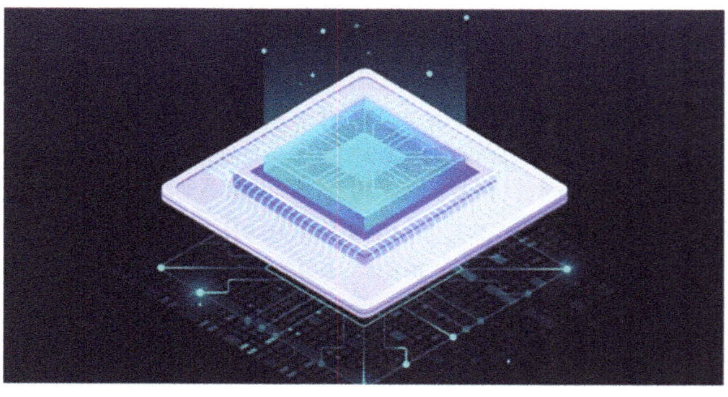

Figure 1.3. Illustration of the chips of tomorrow—DARPA's NGMM program [15]. (Image courtesy of Defense Advanced Research Projects Agency.)

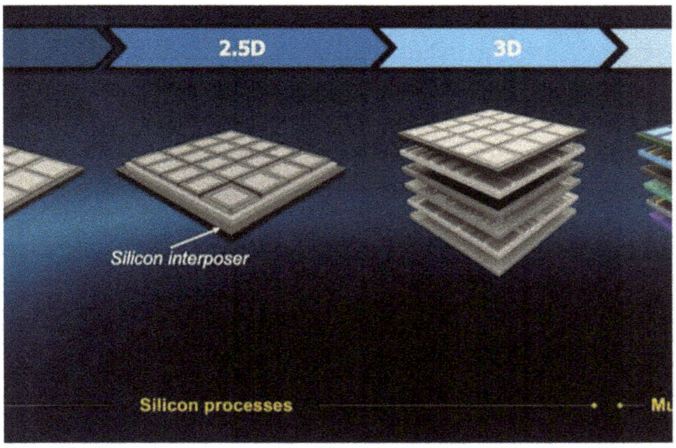

Figure 1.4. Schematic of 3DHI chips [16]. (Image courtesy of Defense Advanced Research Projects Agency.)

the growth of US-based manufacture of three-dimensional heterogeneous integration (3DHI) chips. Figure 1.4 [16] presents a schematic of the chips.

1.8 Summary

Various properties of silicon are summarized in this chapter. Recent initiatives that promote silicon technology include the rapid prototyping and manufacture of 3DHI chips, transistors and ICs that are required to process large amounts of data for artificial intelligence/machine learning (AI/ML) workloads.

References

[1] Burkacky O, Mancini M, Patel M, Poltronieri G and Roundtree T 2024 Exploring new regions: The greenfield opportunity in semiconductors *McKinsey & Company* https://mckinsey.com/industries/semiconductors/our-insights/exploring-new-regions-the-greenfield-opportunity-in-semiconductors (Accessed: 10 December 2024)

[2] Siffert P and Krimmel E 2004 *Silicon: Evolution and Future of a Technology* (Berlin: Springer)

[3] Uddin W *et al* 2022 Silicon-based qubit technology: progress and future prospects *Bull. Mater. Sci.* **45** 46

[4] Tilli M and Haapalinna A 2015 Properties of silicon *Handbook of Silicon Based MEMS Materials and Technologies* 2nd edn ed M Tilli *et al* (Boston, MA: William Andrew) ch 1 pp 3–17

[5] Silicon *Wikipedia* https://en.wikipedia.org/wiki/Silicon

[6] Mills B 2007 File:Silicon-unit-cell-3D-balls.png *Wikimedia Commons* https://commons.wikimedia.org/wiki/File:Silicon-unit-cell-3D-balls.png#filelinks

[7] Lang W 1996 Silicon microstructuring technology *Mater. Sci. Eng.* R **17** 1–55

[8] Balarin M 1988 *Properties of Silicon* EMIS Datareviews Series No. 4 **vol 24** (London: INSPEC, The Institution of Electrical Engineering)

[9] Büttgenbach S *et al* 2020 Piezoresistive pressure sensors *Case Studies in Micromechatronics* (Berlin: Springer) pp 21–85

[10] Honsberg C and Bowden S n.d. General properties of silicon *PV Education* https://pveducation.org/pvcdrom/materials/general-properties-of-silicon

[11] Tilli M and Haapalinna A 2020 Properties of silicon *Handbook of Silicon Based MEMS Materials and Technologies* 3rd edn ed M Tilli *et al* (Amsterdam: Elsevier) ch 1 pp 3–17

[12] Ieong M *et al* 2004 Silicon device scaling to the sub-10-nm regime *Science* **306** 2057–60

[13] Tseng L-T *et al* 2023 Resistless EUV lithography: photon-induced oxide patterning on silicon *Sci. Adv.* **9** eadf5997

[14] Mukesh S and Zhang J 2022 A review of the gate-all-around nanosheet FET process opportunities *Electronics* **11** 3589

[15] DARPA brings next-gen microelectronics manufacturing closer to reality *DARPA* https://www.darpa.mil/news/2024/next-gen-microelectronics-manufacturing (Accessed: 10 December 2024)

[16] DARPA eyes complex 3D microelectronics *SIGNAL* https://afcea.org/signal-media/darpa-eyes-complex-3d-microelectronics (Accessed: 10 December 2024)

Chapter 2

Silicon—types

A concise introduction to the various forms of silicon is presented in this chapter. The associated properties and applications are described briefly.

2.1 Introduction

Silicon (Si) is a versatile material that is widely used in microelectronics, photonics, and energy-related industries. Silicon can be classified based on its purity, structure, and intended use [1]. Each type of silicon, introduced in this chapter, serves unique purposes depending on the required properties, cost constraints, and application domains. Silicon is broadly classified into two types; crystalline silicon (c-Si) and amorphous silicon (a-Si). Crystalline silicon refers to the various crystalline forms of silicon, which include the following: monocrystalline silicon (mono-Si, c-Si), multicrystalline silicon (mc-Si), and polycrystalline silicon (pc-Si) [2]. The other forms of silicon such as amorphous silicon (a-Si), microcrystalline silicon (μc-Si), and nanocrystalline silicon (n-Si) are not categorized as crystalline silicon; they are referred as thin films and are used extensively in thin-film solar cells, liquid crystal displays (LCDs), and flexible electronics [1, 3]. The following section provides a brief description of c-Si, mc-Si, pc-Si, a-Si, μc-Si, and n-Si, along with porous silicon (PSi).

2.2 Crystalline silicon

2.2.1 Monocrystalline silicon

In microelectronics, crystalline silicon (c-Si) usually refers to monocrystalline silicon (mono-Si). It has a uniform crystal structure; it is a highly pure form of crystalline silicon and is extensively used in the manufacture of integrated circuits (ICs) and solar photovoltaic (PV) cells. Mono-Si consists of a silicon crystal lattice in which the entire solid is continuous and free of any grain boundaries. Figure 2.1(a) depicts a schematic representation of the bonding configuration of c-Si. It can be made

doi:10.1088/978-0-7503-3247-7ch2

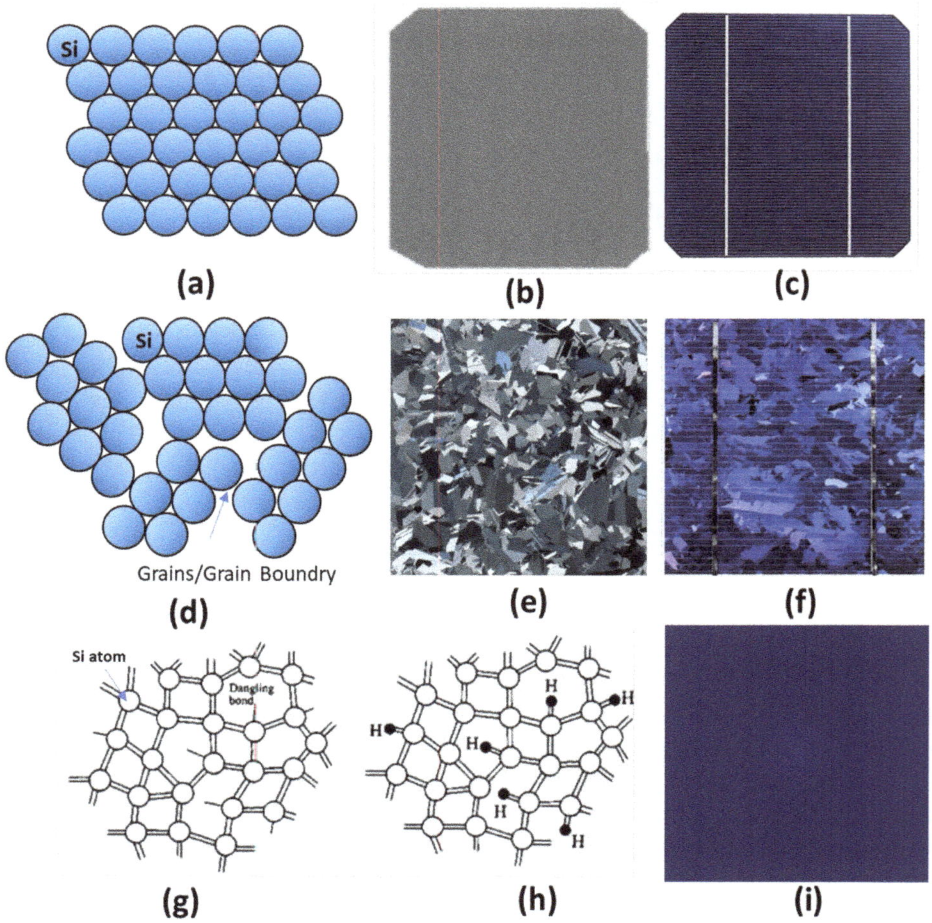

Figure 2.1. (a) The regular arrangement of silicon atoms in monocrystalline silicon. (b) Monocrystalline silicon wafer. (c) Monocrystalline silicon solar cell. (d) Structure of multicrystalline/polycrystalline silicon, showing the boundaries between grains. (e) Textured multicrystalline silicon wafer. (f) Solar cell fabricated from a multicrystalline silicon wafer. (g) Schematic illustration of the structure of amorphous silicon (a-Si) and (h) hydrogenated amorphous silicon (a-Si:H); (i) amorphous silicon solar cell. ((e) Reproduced from [20]. The [Multicrystallinewafer 0001] image has been obtained by the author(s) from the Wikimedia website where it was made available by [Kuebi] under a CC BY-SA 3.0 licence. It is included within this article on that basis. It is attributed to [Kuebi]. Panels (g) and (h) reproduced from [21]. CC BY-SA 4.0.)

either highly pure or doped with appropriate dopants such as boron, arsenic or phosphorous to change its semiconducting properties. Crystalline Si is anisotropic; it is grown from a single, continuous silicon crystal lattice mainly by using Czochralski [4] and float zone [5] techniques. Crystalline Si can be processed into wafers up to 450 mm in diameter. The current industry standard wafer is 12 inches (300 mm) [6]. A silicon wafer has a uniform dark gray color (figure 2.1(b)) whereas a silicon solar cell has deep bluish color (after the deposition of the antireflection coating), as shown in figure 2.1(c).

Mono-Si solar cell efficiencies range from 20% to 26%, depending on the technology [7]. Monocrystalline silicon based solar panels tend to last longer ~25–30 years and perform better in low-light conditions compared to other crystalline silicon types. Although c-Si solar cells are expensive to produce, compared to polycrystalline and other thin-film based PVs, their high efficiency and sleek aesthetics make them the preferred choice in situations where performance or design is a priority. Thus, mono-Si is the gold standard for high-performance PV and semiconductor applications, providing unmatched efficiency and durability at a premium cost. Its pivotal role in modern semiconductor technology ensures its continued significance in renewable energy and electronics.

2.2.2 Multicrystalline silicon

Multicrystalline silicon (mc-Si) is a widely used material in the production of solar cells. It is distinct from c-Si due to its structure and fabrication process. Multicrystalline silicon consists of multiple small silicon crystals, usually with a grain size of ~1 mm–10 cm, separated by grain boundaries, as shown in figure 2.1 (d). These boundaries between separate crystallites induce internal strain in mc-Si and are often seen as a patchwork texture on the surface, giving mc-Si solar cells a distinctive appearance (as can be seen in figure 2.1(e)). Grain boundaries act as sites for impurities and dislocations and serve as recombination centers, reducing carrier lifetime and efficiency [8]. The scattering of phonons and charge carriers on grain boundaries results in reduced thermal and electrical conduction in mc-Si as compared to c-Si [8]. Multicrystalline silicon is more cost-effective to produce than c-Si due to its simpler manufacturing process; it is produced by directional solidification in a quartz crucible [9]. Solar conversion efficiencies of commercial mc-Si cells are in the range of ~15%–20% [7]. The square shape of mc-Si matches better with solar panel designs, reducing wasted space compared to round c-Si wafers (figure 2.1(f)). Although mc-Si solar cells account for about 50% of commercial products, mc-Si technology is being phased out in many markets across the world, due to significant advancements in c-Si PV technology (for example, passivated emitter and rear cell or passivated emitter and rear contact (PERC) cells, and tunnel oxide passivated contact (TOPCon) cells) combined with a significant reduction in its costs [10]. In addition, mc-Si cells are slightly less effective in low-light conditions compared to c-Si [11]. Nonetheless, they are still used in applications where cost is a primary factor over efficiency.

2.2.3 Polycrystalline silicon

Polycrystalline silicon (pc-Si) consists of grains (single crystals) of no preferred orientation and is used broadly in PV applications. Polycrystalline silicon is composed of numerous smaller silicon grains, usually with a crystal size of 1 μm–1 mm, which are smaller than those of mc-Si [1]. It is less expensive to produce pc-Si than c-Si and this material offers a good balance of performance and cost. The solar cell produced from poly-Si has a similar bluish color to the mc-Si solar cell with a speckled or mosaic-like texture due to its grain structure, as shown in figure 2.1(f).

Polycrystalline silicon is produced from metallurgical grade Si using a chemical purification process. High-quality polysilicon can be produced using two primary methods: fluidized bed reactor (FBR) [12] technology and the Siemens process [12, 13]. High-purity pc-Si is used as the base material for the manufacture of silicon wafers and the substrate for integrated circuits (ICs), and is used in thin-film transistors (TFTs), LCDs, and other flat-panel display technologies. Polycrystalline silicon solar cells have efficiency of ~15%–20% [7], lower than that of c-Si solar cells due to grain boundaries causing electron scattering and recombination [14]. However, it has lower production costs than c-Si due to simpler manufacturing processes and less material wastage. It is cost-effective and an affordable option for residential and utility-scale solar installations. Solar cells based on pc-Si are slightly more prone to efficiency loss at high temperatures compared to c-Si solar cells. Nevertheless, poly-Si is an excellent choice where cost savings and reliability are more critical than maximum efficiency. The use of pc-Si strikes a balance between affordability and performance, making it popular for a broad range of solar applications.

2.3 Amorphous silicon

Amorphous silicon (a-Si) is a non-crystalline form of silicon with a disordered atomic structure. It forms a continuous random network and all atoms in a-Si are not four-fold coordinated, as shown in figure 2.1(g). The absence of crystalline order leads to a high density of dangling bonds (figure 2.1(g)), which can trap charge carriers. Hydrogenation, i.e. forming hydrogenated amorphous silicon (a-Si:H), as shown in figure 2.1(h), has sufficiently lower defects and is used in the manufacture of devices such as solar cells. Hydrogenated amorphous silicon is a direct-bandgap semiconductor (~1.5 to ~1.9 eV). This makes it better at absorbing visible light, which is beneficial for thin-film PV. The variation in the bandgap is due to the preparation conditions, but the most important parameter is the hydrogen content. It can be deposited as thin films on a variety of substrates including glass, plastic, metal, and flexible substrates at low temperatures (~75 °C) [15]. Amorphous silicon is commonly used in the fabrication of thin-film solar cells, roll-to-roll processing, display technology, and electronics [16]. Amorphous silicon is doped, just like c-Si, to form p-type and n-type layers, which are used to form various electronic and photonic devices. Amorphous silicon is deposited using a range of physical and chemical methods, including plasma-enhanced chemical vapor deposition (PECVD), magnetron evaporation, and extremely high frequency CVD [15, 17]. However, PECVD is the most prevalent method for the deposition of a-Si thin films. Amorphous silicon performs well under low light and diffuse conditions, making it suitable for applications such as calculators, small gadgets, and building-integrated photovoltaics (BIPV) [18]. Although a-Si solar cells have achieved lower efficiency (~6%–12%) compared to that of c-Si solar cells (~18%–24%) [7], they are very cost-effective, lightweight, and durable. They are also used in TFTs for LCD screens as well as in photodetectors and image sensors due to their sensitivity to light. Despite these advantages, a-Si solar cells suffer from the Staebler–Wronski effect (SWE) [19], due to which prolonged exposure to sunlight reduces their efficiency over time and leads to a shorter lifespan, especially in outdoor applications where degradation can be more significant.

2.4 Microcrystalline silicon

Microcrystalline silicon is a form of silicon used in thin-film solar cells, transistors, organic light-emitting diodes (OLEDs), LCDs, and other semiconductor applications. It is related to amorphous silicon but has distinct properties due to its mixed-phase structure. Microcrystalline silicon (μc-Si) consists of small silicon crystals embedded in an amorphous silicon matrix. The separation of the nano- and microcrystalline phases inside the amorphous matrix is shown in the schematics in figures 2.2(a) and 2.2(b), respectively. Thin films with silicon crystallites larger than 10 nm are referred to as microcrystalline [22] (figure 2.2(b)). Electronic transport in μc-Si layers can be obstructed by the many grain boundaries, when not passivated by hydrogen (H_2) [23]. Hydrogenated microcrystalline silicon (μc-Si:H) has high stability against light and electrical stress [23]. It has an indirect bandgap ($\sim$1.1 eV) and can be used as an absorber layer in tandem or multi-junction solar cells, often paired with amorphous silicon. Tandem cells exploit its higher absorption in the infrared spectrum while amorphous silicon absorbs visible light. Thin-film solar cells with microcrystalline silicon have achieved efficiencies between 8% and 12% [7] (less efficient compared to crystalline silicon solar cells) in commercial applications, although laboratory cells have reached higher efficiencies. It is less prone to the SWE than amorphous silicon [24]. Microcrystalline silicon is generally deposited using PECVD or hot-wire CVD (HWCVD) [25]. Common precursor gases are silane (SiH_4) diluted with hydrogen (H_2). It can be deposited at low temperatures (200 °C–400 °C), making it compatible with flexible substrates such as plastic or glass. The degree of crystallinity can vary depending on the deposition conditions and influences its electrical and optical properties. The agglomeration of fine silicon

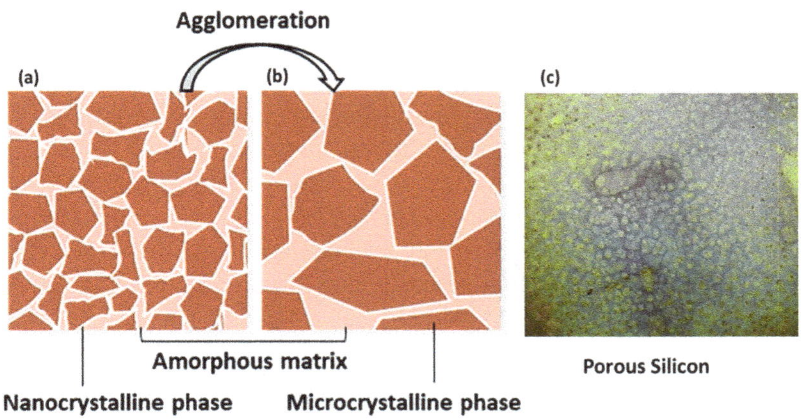

Figure 2.2. Schematic illustration of (a) nano and (b) microcrystalline silicon within the amorphous silicon matrix. (c) Stains produced on the silicon surface during electroless chemical etching. ((a) and (b) Reproduced from [26]. CC BY 3.0. (c) Reproduced from [31]. CC BY 3.0.)

nanocrystallites results in the creation of microcrystallites, depending on the process conditions [26]. The slow deposition process increases production costs and the presence of grain boundaries and defects can limit carrier mobility and lifetime. A precise control of the layer thickness is required to balance light absorption and electrical conductivity.

2.5 Nanocrystalline silicon

Nanocrystalline silicon (nc-Si), also known as μc-Si, is a form of porous silicon. It consists of nanometer-sized crystalline grains embedded in an amorphous silicon matrix. This material combines the properties of both crystalline and amorphous silicon, making it highly versatile for applications in electronics, photovoltaics, and optoelectronics. The films are regarded as nanocrystalline if the crystallites are densely occupied and are only a few nanometers in size, as shown in figure 2.2(a) [22]. It has higher electrical conductivity, improved carrier mobility and stability, reduced light-induced degradation (SWE), and efficient light trapping compared to amorphous silicon due to the presence of silicon crystallites. nc-Si is easier to fabricate than poly-Si [27] and the commonly used method is PECVD [26]. Growth conditions such as the temperature, gas mixture, and plasma power influence the crystallinity and quality of nc-Si. Nanocrystalline silicon has applications in large-area solar cells, photodetectors, and sensors, display technologies such as LCD and OLED screens, and oligonucleotide chips [26, 28]. It is often used as the intrinsic layer in multi-junction solar cells and incorporated into tandem solar cells (e.g. with amorphous silicon) to capture a broader range of wavelengths in the solar spectrum. Despite its several advantages, the control of crystallinity during deposition is complex and the material can exhibit structural defects at grain boundaries, affecting the device performance. In addition, the cell efficiency remains lower than that of traditional c-Si based solar cells.

2.6 Porous silicon

Porous silicon (PSi) was accidentally discovered by Uhlir at Bell Laboratories in 1956 [29]. Porous silicon is a form of silicon characterized by a network of interconnected pores, giving it a highly porous structure. This material is versatile and has unique optical, electrical, and mechanical properties due to its high surface area and nanoscale features [30, 31]. It is composed of a silicon matrix with nanometer to micrometer sized pores. The porosity ranges from a few percent to over 90%, depending on the preparation methods and conditions [31]. PSi exhibits photoluminescence and electroluminescence due to quantum confinement effects in the nanoscale silicon structures. A tunable bandgap can be achieved by adjusting the pore size and structure. The fragility of PSi increases with higher porosity and it has high chemical reactivity due to a large surface area. The PSi surface can be functionalized for specific applications. Porous silicon can be prepared by various processes such as anodization, stain etching, and bottom-up synthesis [32]. One such process involves a silicon wafer submerged in a solution (usually hydrofluoric acid) and subjected to anodic bias. Pore size and distribution depend on the etching conditions (current density, electrolyte composition, and duration). A chemical method, using a mixture of hydrofluoric acid and an oxidizing agent, produces porous silicon without an external current [32]. Figure 2.2(c) depicts the stains formed on the surface of the silicon wafer using electroless chemical etching [31]. Dry etching processes such as plasma or reactive ion etching create porous structures. In the laser-induced technique, a laser is used to locally enhance the etching processes. PSi is used in photonic crystals, waveguides, light-emitting devices, infrared sensors, and detectors. It can be used for biosensing applications due to its sensitivity to changes in its environment [33]. Furthermore, the bio-compatibility and biodegradability of PSi make it ideal for medical applications [34]. It is also used as an anode material in lithium-ion batteries, leveraging its high surface area for enhanced capacity [35]. In addition, it is used as a sacrificial layer in micro-electromechanical systems (MEMS) for creating microstructures [34]. In solar cells, PSi acts as an antireflective coating to improve light absorption [36]. PSi is a fragile material, especially at high porosities. Therefore, controlling uniformity and pore size distribution can be challenging and the stability under environmental conditions, such as oxidation, requires careful handling.

2.7 Summary

A brief overview of the various structures of silicon is summarized in this chapter. The relevant processes and applications are highlighted.

References

[1] Honsberg C and Bowden S 2019 Types of silicon *PV Education* https://pveducation.org/pvcdrom/manufacturing-si-cells/types-of-silicon

[2] Hanif M A *et al* 2022 Solar thermal energy and photovoltaic systems *Renewable and Alternative Energy Resources* ed M A Hanif *et al* (New York: Academic) ch 4 pp 171–261

[3] Avrutin V, Izyumskaya N and Morkoç H 2011 Semiconductor solar cells: recent progress in terrestrial applications *Superlattices Microstruct.* **49** 337–64

[4] Zulehner W 1983 Czochralski growth of silicon *J. Cryst. Growth* **65** 189–213

[5] Dietze W, Keller W and Mühlbauer A 1981 Float-zone grown silicon *Silicon* (Berlin: Springer)

[6] WaferPro 2024 The gradual growth of silicon wafer sizes: an evolutionary history *WaferPro* https://waferpro.com/the-gradual-growth-of-silicon-wafer-sizes-an-evolutionary-history/

[7] Best research-cell efficiency chart *NREL* https://www.nrel.gov/pv/cell-efficiency

[8] Stokkan G, Song A and Ryningen B 2018 Investigation of the grain boundary character and dislocation density of different types of high performance multicrystalline silicon *Crystals* **8** 341

[9] Yang Y M *et al* 2015 Development of high-performance multicrystalline silicon for photovoltaic industry *Prog. Photovolt. Res. Appl.* **23** 340–51

[10] Bošnjaković M 2024 Advance of sustainable energy materials: technology trends for silicon-based photovoltaic cells *Sustainability* **16** 7962

[11] Reich N H *et al* 2009 Crystalline silicon cell performance at low light intensities *Sol. Energy Mater. Sol. Cells* **93** 1471–81

[12] Bernreuter J 2020 Polysilicon production processes: Siemens process *Bernreuter Research* https://bernreuter.com/polysilicon/production-processes/ (Accessed: 12 December 2024)

[13] Yan D 2019 Siemens process *Handbook of Photovoltaic Silicon* ed D Yang (Berlin: Springer) pp 37–68
[14] Greuter F and Blatter G 1990 Electrical properties of grain boundaries in polycrystalline compound semiconductors *Semicond. Sci. Technol.* **5** 111
[15] Chittick R C, Alexander J H and Sterling H F 1969 The preparation and properties of amorphous silicon *J. Electrochem. Soc.* **116** 77
[16] Street R A 2000 *Technology and Applications of Amorphous Silicon* (Berlin: Springer)
[17] Robertson J 2000 Deposition mechanism of hydrogenated amorphous silicon *J. Appl. Phys.* **87** 2608–17
[18] Li X *et al* 2021 Review and perspective of materials for flexible solar cells *Mater. Rep.: Energy* **1** 100001
[19] Zeman M *et al* 1997 Computer modelling of current matching in a-Si:H/a-Si:H tandem solar cells on textured TCO substrates *Sol. Energy Mater. Sol. Cells* **46** 81–99
[20] Kübelbeck A 2007 File:Multicrystallinewafer 0001.jpg *Wikimedia Commons* https://commons.wikimedia.org/wiki/File:Multicrystallinewafer_0001.jpg (Accessed: 7 May 2025)
[21] Amorphous semiconductors *LibreTexts Chemistry* https://chem.libretexts.org/@go/page/189557
[22] Roca i Cabarrocas P 2004 New approaches for the production of nano-, micro-, and polycrystalline silicon thin films *Phys. Status Solidi* c **1** 1115–30
[23] Shah A V *et al* 2003 Material and solar cell research in microcrystalline silicon *Sol. Energy Mater. Sol. Cells* **78** 469–91
[24] Günes M *et al* 2012 Metastability effects in hydrogenated microcrystalline silicon thin films investigated by the dual beam photoconductivity method *J. Non-Cryst. Solids* **358** 2074–7
[25] Klein S, Repmann T and Brammer T 2004 Microcrystalline silicon films and solar cells deposited by PECVD and HWCVD *Sol. Energy* **77** 893–908
[26] Sharma M, Panigrahi J and Komarala V K 2021 Nanocrystalline silicon thin film growth and application for silicon heterojunction solar cells: a short review *Nanoscale Adv.* **3** 3373–83
[27] Bauza M *et al* 2009 Light-induced metastability in thin nanocrystalline silicon films *Philos. Mag.* **89** 2531–9
[28] Wei D, Xu S and Levchenko I 2019 Nanocrystalline silicon and solar cells *Handbook of Photovoltaic Silicon* ed D Yang (Berlin: Springer) pp 793–841
[29] Uhlir A Jr and Uhlir I 2005 Historical perspective on the discovery of porous silicon *Phys. Status Solidi* c **2** 3185–7
[30] Theiß W 1997 Optical properties of porous silicon *Surf. Sci. Rep.* **29** 91–192
[31] Farshid K 2018 Porous silicon *Porosity* ed G Taher Hcine (Rijeka: IntechOpen) ch 1
[32] Kuntyi O, Zozulya G and Shepida M 2022 Porous silicon formation by electrochemical etching *Adv. Mater. Sci. Eng.* **2022** 1482877
[33] Moretta R *et al* 2021 Porous silicon optical devices: recent advances in biosensing applications *Sensors* **21** 1336
[34] Sailor M J 2012 *Porous Silicon in Practice: Preparation, Characterization and Applications* (New York: Wiley)
[35] Jia H *et al* 2020 Hierarchical porous silicon structures with extraordinary mechanical strength as high-performance lithium-ion battery anodes *Nat. Commun.* **11** 1474
[36] Singh P, Sharma S N and Ravindra N 2010 Applications of porous silicon thin films in solar cells and biosensors *JOM* **62** 15–24

IOP Publishing

Black Silicon
Technology, opportunities and challenges
N M Ravindra and Priyanka Singh

Chapter 3

Black silicon

The chapter provides a basic description of black silicon (BSi). The surface texturization techniques and some related recent developments in the methods of formation of BSi are discussed. An overview of the antireflection coating (ARC) principle for BSi solar cell applications is presented. The modeling techniques that have been described in the literature are explored briefly, along with applications of current interest.

3.1 Introduction

Black silicon (BSi) is a surface modification of silicon (Si) and is chemically identical to Si. The surface of BSi consists of a top layer of nano- or fine microstructures (a textured surface) that are produced by etching [1], as first reported by Jansen *et al* [2]. As shown in figure 3.1, etching results in the silicon wafers appearing black (figure 3.1(b)), as compared to planar silicon wafers, which are typically silver-gray (figure 3.1(a)) [3]. The nanoscale structure of BSi exhibits an incredibly low reflectivity of ~0% (generally <5%), while traditional silicon reflects roughly 30% of the incident sunlight [1, 2, 4] as depicted in figure 3.2. The most widely used method for fabricating BSi layers is reactive ion etching (RIE) in a gas mixture of sulfur hexafluoride (SF_6) and oxygen (O_2) [4]. The nanostructures of BSi allow the material to absorb a broader spectrum of light, including infrared. As a result, BSi has the potential to replace conventional antireflection coating (ARC) in photovoltaic (PV) solar cells. Furthermore, the nanostructured surface of BSi exhibits excellent absorption over a wide spectral range (250–2500 nm), which not only makes it ideal as an ARC [5, 6] but also makes it applicable in photodetectors [7], photodiodes [8], and gas sensors [9]. BSi is being explored widely for improving the efficiency of silicon-based solar cells by capturing more light, especially in diffuse lighting conditions. Its high sensitivity to light makes it suitable for applications in photodetectors that detect low-intensity or broad-spectrum light. It is also used in

doi:10.1088/978-0-7503-3247-7ch3

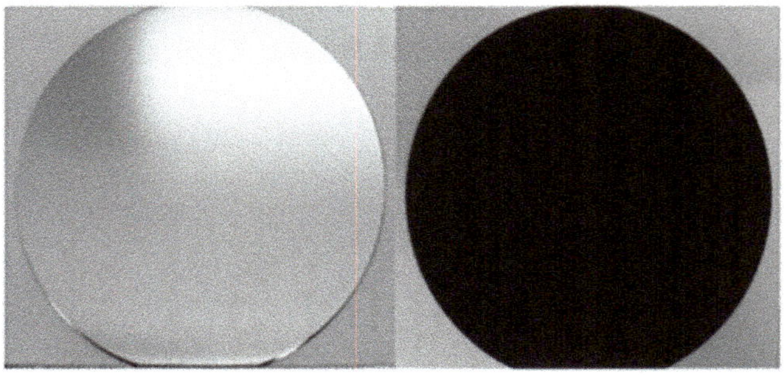

Figure 3.1. Photograph of (a) a bare polished silicon wafer and (b) a BSi surface after etching. (Reproduced with permission from [3]. Copyright 2023 Royal Society of Chemistry.)

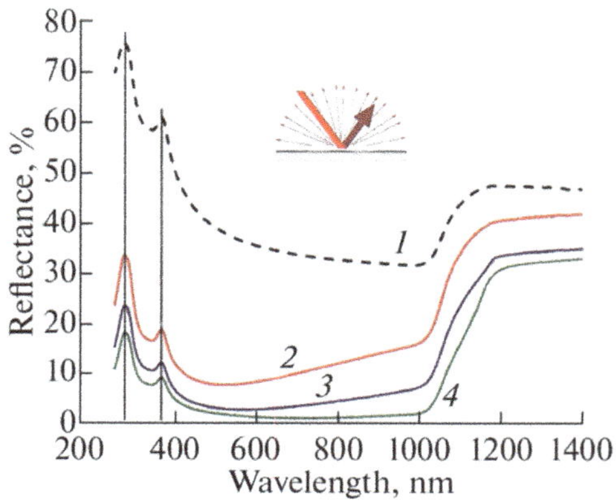

Figure 3.2. Reflectance spectra of (1) a silicon substrate and with a BSi layer, formed with etching durations of (2) 3 min, (3) 6 min, and (4) 10 min. (Reproduced with permission from [4]. Copyright 2024 Springer Nature.)

scientific instruments such as optical sensors where reduced reflectivity and high light absorption are critical.

3.2 Surface texturing

Surface texturing of substrates refers to the modification of a material's surface to improve its functional properties [10]. Surface texturing involves creating micro- or nanoscale patterns on a substrate to alter its physical or optical characteristics [10]. These textures can range from simple grooves or dimples to complex hierarchical structures. In photovoltaics, optoelectronics, and other related applications, surface texturing is often used to enhance light absorption, reduce reflection, or modify surface chemistry [1]. In order to improve the efficiency of solar cells, light trapping is a fundamental feature, and texturing increases light absorption due to an increase in the optical path length by refraction, as shown in figure 3.3(a).

Surface texturing also improves internal reflection, which reduces the amount of light that escapes from the material during interactions with the incident light. In this

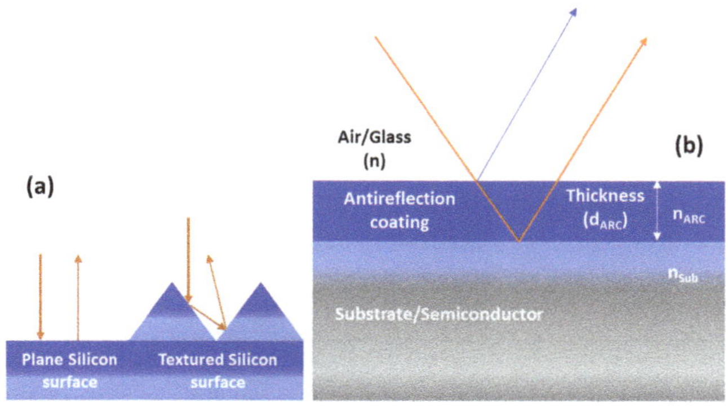

Figure 3.3. (a) Representation of increased light trapping in a textured surface. (b) Schematic representation of anti-reflective coating.

section, the basic principle of ARC in solar cells will be explained briefly for BSi applications, followed by a brief introduction to texturization methods. Texturization techniques will be addressed in detail in chapter 4.

3.2.1 Antireflection coating

BSi is primarily attractive to PV industry because it can be used as an inexpensive, efficient ARC for commercial solar cell applications and can potentially replace traditional ARC. ARC is an optical coating which is applied on the surface of lenses and other optical devices to minimize reflection. Figure 3.3(b) shows a schematic illustration of an ARC. ARCs generally consist of a thin layer of dielectric material, with a specific chosen thickness (d), so that the interference effects in the coating cause the wave reflected from the ARC surface to be out of phase with the wave reflected from the semiconductor surface. These out-of-phase reflected waves cause destructive interference, resulting in zero net reflected energy. For a quarter wavelength ARC of a material with a refractive index n and light incident on the coating with a free-space wavelength λ, the thickness d_{ARC} that causes minimal reflection is computed as

$$d_{ARC} = \frac{4\pi}{n\lambda}. \tag{3.1}$$

Reflection is further reduced if the refractive index of the ARC is the geometric mean of the refractive indices of the materials on either side, that is, glass or air and the semiconductor. This is expressed as

$$n_{ARC} = \sqrt{(n \times n_{semi})}, \tag{3.2}$$

where, n_{ARC}, n, and n_{semi} are the refractive indices of the ARC, glass (or air), and semiconductor, respectively.

The use of ARC results in a very low reflectance (R_{min}) at a specific wavelength (λ_{min}). This effect reduces the reflectance in a wavelength domain centered on λ_{min}, which for PV applications must lie between 650 and 700 nm (visible range), where

Table 3.1. Performance data comparisons for different photovoltaic cells. Reproduced from [12].

Sample/method	V_{oc} (V)	I_{sc} (A)	FF (%)	η(%)
BSi (ICP-RIE) [12]	0.622	0.364	82.2	23.3
BSi (DRIE) [13]	0.665	0.422	78.7	22.1
BSi (RIE) [14]	0.620	0.350	80.0	17.4
BSi (PIII) [15]	0.621	0.320	78.3	15.7
BSi (MACE) [16]	0.628	0.373	76.7	17.9
$SiO_2/WO_3/ZnO$ (80/20) [17]	0.437	0.207	48.5	5.48

the photon energy in the solar spectrum peaks [11]. Like most materials, crystalline Si has a wavelength-dependent refractive index (3.84 at 650 nm, 3.76 at 700 nm). Accordingly, the refractive index of an optimal ARC material should be equal to 1.96 at 650 nm and 1.94 at 700 nm [11]. As a result of equations (3.1) and (3.2), an ARC thickness of 83 nm (90 nm) is required to achieve zero net reflectance at 650 nm (700 nm).

In a recent work, BSi has been implemented as an ARC in solar cells. BSi layers were fabricated using inductively coupled plasma reactive ion etching (ICP-RIE) technology [12]. In table 3.1, a comparison of the results of the photovoltaic performance parameters, namely, open circuit voltage (V_{oc}), short circuit current (I_{sc}), fill factor (FF), and efficiency (η) of the solar cell using the ICP-RIE method [12] and solar cells using different fabrication methodologies, is presented.

It can be noted in table 3.1 that the fill factor was higher (82.2%) for cells fabricated using ICP-RIE [12] than other solar cell manufacturing methods. Additionally, with optimized nanostructures and efficient passivation, BSi solar cells attained an efficiency of 23.3%.

3.2.2 Texturization methods

Surface texturization methods can generally be classified into three broad categories: wet etching, dry etching, and laser based etching [18, 19]. An illustration of the fabrication processes, properties, and associated applications of BSi is shown in figure 3.4 [20].

Wet chemical texturization techniques include isotropic etching, anisotropic chemical etching, metal-assisted chemical etching (MACE), and electrochemical etching (ECE) [18]. Generally, aqueous alkaline etching produces upright pyramids on the monocrystalline silicon surface [21]; acidic etching creates isotropic structures or random textures. In a study, experiments were performed to produce uniform microstructure upright pyramids on Si with alkaline etching with different Triton-X concentrations ($<$100 ppm) [22]. An SEM image showing an average height of 11 μm, etched with 30 ppm Triton-X, is shown in figure 3.5(a), which shows a successful wide-area coverage and a uniform growth of pyramidal microstructures. For the first time, a recent study published in the literature shows the fabrication of inverted pyramid structures on monocrystalline silicon surfaces using ethylene glycol

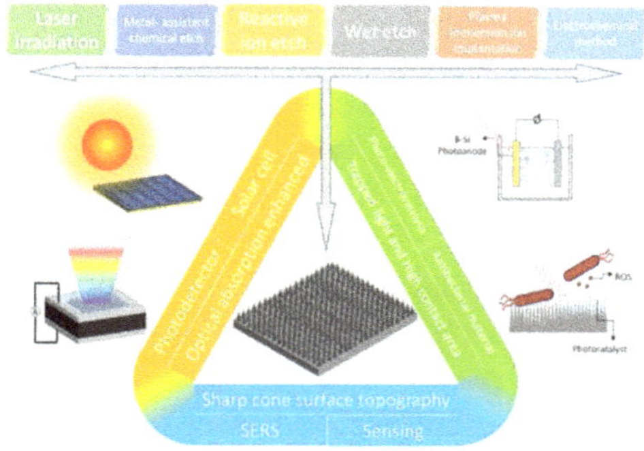

Figure 3.4. An illustration of the fabrication processes and properties, and relevant applications of BSi. (Reproduced from [20]. Copyright MDPI 2021, CC BY 4.0.)

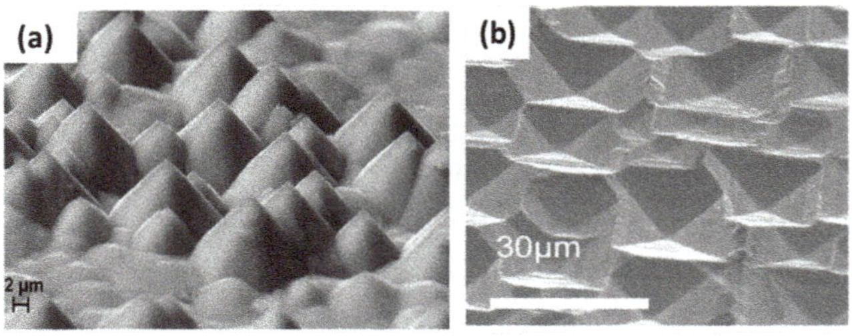

Figure 3.5. (a) Top view of an SEM micrograph of silicon etched with 30 wt% KOH, 17 vol% methanol, and 30 ppm Triton-X. (b) SEM image with a 45 ° oblique view of inverted pyramid structure prepared by etching p-type silicon wafers. ((a) Reproduced with permission from [22]. Copyright 2020 Elsevier. (b) Reproduced with permission from [23]. Copyright 2024 Elsevier.)

butyl ether (EGBE) to regulate aqueous anisotropic alkaline etching [23]. Figure 3.5 (b) shows SEM images of the inverted pyramid cross-sections prepared by alkaline texturization.

The etching process is simple and cost-effective as well as suitable for large-scale production. However, it has limited control over the uniformity of the nano-structure. Environmental and safety concerns are also due to chemical waste. In the MACE process [24], a thin metal catalyst layer (e.g. silver, gold, or platinum) is deposited on the silicon surface and a mixture of HF and H_2O_2 is used to create nanostructures. This method produces highly uniform nanostructures such as vertical nanowires or pores and is compatible with various scales and complex geometries. However, it requires precise control of catalyst deposition, and causes potential contamination from metal residues. In electrochemical etching, silicon is immersed in an electrolyte solution, and an electric current is applied to produce pores or nanostructures. The porosity and structure size of the BSi surface is based on current density and solution parameters. It produces uniform texturing over large areas and is cost-effective for controlled porosity. However, it has limited scalability and requires precise current and solution control.

Dry etching texturization methods for the formation of BSi include the following: RIE [2], deep reactive ion etching (DRIE) or the Bosch™ process [25], and plasma etching [26]. RIE combines plasma and chemical etching to create highly controlled textures [27]. The RIE plasma etching system is used with reactive gases (e.g. SF_6, O_2) to etch the silicon surface which generates random or patterned nanostructures, depending on the mask [1, 2, 27]. RIE is a highly precise method and produces uniform structure and is compatible with microfabrication processes. However, it is expensive and requires complex equipment and has limited scalability for large wafers.

Laser based methods utilize femtosecond laser irradiation [28, 29]. The femtosecond laser irradiation technique uses femtosecond pulsed lasers to ablate the silicon surface to create micro- and nanostructures [28]. It can be performed in the presence of reactive gases such as sulfur hexafluoride (SF_6) to dope the silicon surface with sulfur. It is highly flexible in structure design. It can induce surface doping, enhancing optical absorption. However, the equipment cost is high and the process consumes significant energy. It is not particularly suitable for large-scale production. Lithographic techniques include photolithography, electron beam lithography, and nanoimprint lithography [29]. Lithography employs masks to pattern nanostructures onto a silicon wafer. This allows precise control over the nanostructure geometry and is scalable for industrial applications. However, it requires high-quality masks and involves multiple fabrication steps.

In addition to the above methods, some hybrid methods combine wet etching with RIE or laser methods to balance cost, scalability, and structural precision, for example, starting with MACE for rough patterning, followed by RIE for finer control [30, 31].

A recent study [24] presents a complete evaluation of the use of MACE to fabricate BSi-based thermal absorbers in thermophotovoltaic (TPV) systems with effective wavelengths ranging from 600 to 2000 nm, corresponding to energy levels of 0.55–1.85 eV. The MACE technique is relevant and useful in TPV systems due to its ability to fabricate silicon-based thermal absorbers. As demonstrated in figure 3.6, when radiation enters the surface of the BSi, it rapidly scatters the radiation towards neighboring nanostructures, and therefore large reflection losses can be counteracted. Because the BSi absorbs the majority of the incoming radiation and only reflects a small portion, it can theoretically function as a thermal absorber in a TPV system [24].

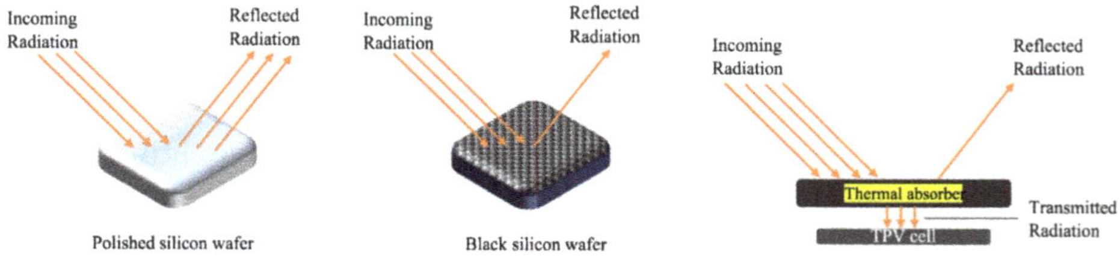

Figure 3.6. Illustration of reflectance reduction in BSi of incoming photons on its surface. (Reproduced from [24]. Copyright 2025 IOP. CC BY 4.0.)

It was discovered that the BSi pyramidal textures that are made by copper-induced cycling reactions exhibited useful antireflection characteristics. Solar cells produced utilizing this process exhibited a conversion efficiency of 10.7% [32]. Additional details of the related experiments can be found in [32].

Figure 3.7 shows the morphological evolution of Cu-induced cycle etching when H_2O_2 concentrations were steadily increased from 0.14 to 0.53 to 0.65 M and subsequently to 0.90 M, respectively. These results demonstrated a substantial link between etching profiles and H_2O_2 concentrations. Si textures were not visible on Si surfaces at low H_2O_2 concentration (0.14 M) after etching, as shown in the SEM cross-sectional and top views in figure 3.7(a) [32]. The addition of H_2O_2 up to 0.53 M significantly affected the etching characteristics. It is also clearly indicated that the etching properties changed substantially from almost isotropic to anisotropic patterns as H_2O_2 concentrations increased. The creation of pyramid-shaped textures is clearly visible in figure 3.7(b). Increasing the concentration of H_2O_2 to 0.65 M enhanced the Si texturization and Cu dissolution. Instead, the morphologies of the etching pores were modified to inverted pyramid structures, as shown in figure 3.7(c). The Cu-induced cycle etching resulted in uneven and rough surfaces at high concentrations of H_2O_2 (0.90 M), as illustrated in figure 3.7(d) [32].

In a study, the effect of argon (Ar) gas additives on the production of BSi has been demonstrated [33]. In particular, this study investigated the cryogenic unmasked etching of BSi using Ar gas additives. The resulting BSi exhibits a low reflectance of 1% ± 0.5% in the 450–1000 nm region on a 100 mm Si wafer surface. Figure 3.8 shows the SEM images of the Si surface etched without Ar (figure 3.8(a)) and with a 5 sccm Ar flow (figure 3.8(b)) [33]. Despite their same height, Si nanowires (SiNWs) have distinct forms. SiNWs etched without Ar have a pointed top shape (figure 3.8(a)), whereas Ar dilution causes them to be cone-shaped (figure 3.8(b)). The conic shape of

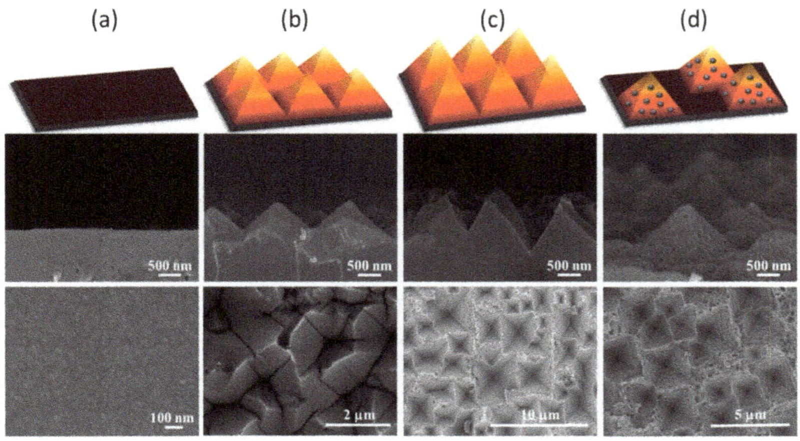

Figure 3.7. Schematic depiction, cross-sectional view, and top view SEM micrographs of BSi textures (from top to bottom) formed with varied concentrations of H_2O_2 during Cu-induced cycle etching: (a) $[H_2O_2] = 0.14$ M, (b) $[H_2O_2] = 0.53$ M, (c) $[H_2O_2] = 0.65$ M, and (d) $[H_2O_2] = 0.90$ M. (Reproduced from [32]. Copyright 2017 Springer Nature. CC BY 4.0.)

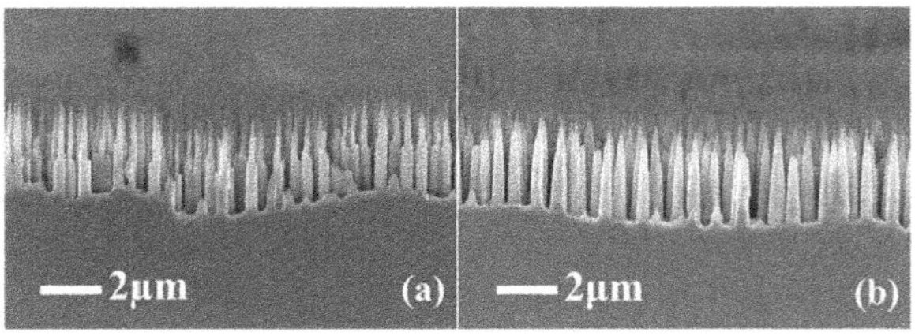

Figure 3.8. SEM micrograph of BSi etched for 200 s without Ar (a) and with 5 sccm Ar (b). (Reproduced from [33]. Copyright 2022 American Chemical Society, CC-BY 4.0.)

SiNWs is significantly favorable for passivation due to further deposition of the passivation layer [33].

3.3 Modeling

Simulations are helpful in learning about the influence of geometry of a nano-textured surface on its optical response. Furthermore, experimental samples are highly variable, but simulations are entirely reproducible, with nanopore sizes and shapes that may be precisely tuned for research needs [34]. Several modeling techniques are available to optimize the ARCs, microscale pyramids, and nanowires [35]. The transfer matrix method (TMM) is a powerful analytical technique for modeling the light interaction with thin films and multilayer structures, and is an effective way to model ARCs [36]. Modeling facilitates determining the amount of incident light that is reflected and transmitted at each interface between any two connected domains [36]. When applied to BSi, TMM can help in understanding and optimizing its optical performance, particularly in photovoltaic applications. For TMM, the nanostructured layer is divided into multiple sub-layers, each characterized by a wavelength-dependent refractive index and thickness. Nevertheless, optical path enhancement through changes in the propagation direction of light by textured surfaces is not included in TMM [35, 36]. Ray-tracing (RT) techniques can be used to represent this, as the technique helps to determine the path of waves through discrete media, considering changes in direction through reflection and refraction at interfaces [35]. RT assumes that light propagates as rays and is particularly effective for modeling structures where the feature size is much larger than the wavelength of light. It is computationally efficient and appropriate for studying the macroscopic optical properties such as reflection, absorption, and light trapping. It works effectively for large scales and is well-suited for modeling the macroscopic light propagation in large BSi surfaces or solar cells. It is capable of managing complex structures and effortlessly integrating random and non-periodic nanostructures. However, RT is incompatible with the more recent nanoscale BSi textures that solar cell manufacturers are using, where surface features are frequently much smaller than the incident wavelength of light [37]. The finite-difference time-domain (FDTD) is another robust numerical approach for simulating the interaction of electromagnetic waves with the complex structures of BSi [38]. The nanostructured surface of BSi is particularly well-suited for FDTD modeling because this approach handles the irregular geometries and intricate light scattering properties that are inherent to these structures. FDTD captures detailed light interaction, including scattering, diffraction, and interference. It is also suitable for irregular and non-periodic nanostructures. It enables wavelength-dependent analysis for broadband applications. However, FDTD does not provide access to the complexities of the simulation geometry that are accessible through the finite element method (FEM) techniques [39]. The FEM is a powerful numerical approach to model and analyse the optical and electrical properties of BSi. The FEM is used to model the reflectance spectra of BSi surfaces. The FEM is particularly suitable for solving Maxwell's equations in complex geometries and material configurations, making it ideal for modeling the complex nanostructures of BSi [40]. Software tools such as Sentaurus TCAD can be used to model the multiscale hybrid structure of BSi by combining RT and TMM [41]. Sentaurus TCAD is a powerful simulation platform that is used widely for modeling and optimizing semiconductor devices, including those incorporating nanostructured materials such as BSi. When using Sentaurus TCAD for BSi, the focus is typically on the simulation of optical properties, electrical behavior, and overall device performance. The effective index technique (EIT) is another widely used approach to simplify the modeling of complex nanostructured surfaces, such as BSi [42]. Instead of explicitly modeling every feature of the nanostructures, EIT replaces the surface with a graded-index layer that approximates the average optical behavior of the nanostructures. This method is computationally efficient and suitable for analysing the optical properties of BSi. In addition to utilizing CAD software, researchers have created a 3D model by scanning the BSi surface to identify peak points to measure the height or depth [43]. Scanning electron microscopy (SEM) or atomic force microscopy (AFM) images can be directly imported for numerical simulation using the FDTD code (Lumerical) [44]. Effective medium theory (EMT) is another alternative mathematical framework used to approximate the optical or electrical behavior of composite materials [45]. For BSi, EMT can model the nanostructured surface as a graded-index material, simplifying simulations by avoiding the explicit modeling of individual nanostructures. It captures optical behavior over a wide range of wavelengths. A study models BSi using TMM and EMT [45]. The model discusses the effects of the BSi geometrical factors, including the structure disorder. The described model is computationally efficient and may be used to predict and optimize BSi performance in a variety of applications such as solar cells, photodetectors, and Raman scattering. Figure 3.9 shows the measured reflectance spectra of BSi, which is compared with the results of the simulation using the proposed model. The TMM model has an excellent agreement with the measured data within 2% [45]. In addition, rigorous coupled wave analysis (RCWA) is a powerful computational method for simulating and analysing the interaction of light with periodic nanostructures. The unique optical properties of BSi, including its high absorption and anti-reflective characteristics, make it an interesting subject for RCWA. In a related study, the reflectance of BSi is simulated with a proprietary RCWA-based tool and compared with experimental results [34]. The findings demonstrate that RIE, followed by isotropic chemical etching can achieve average

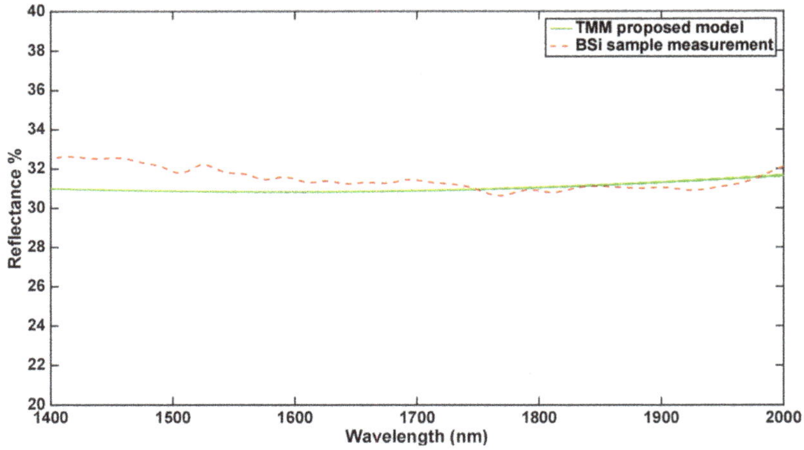

Figure 3.9. The results of the proposed model simulation compared with the reflectance measurement for a BSi sample. (Reproduced from [45]. Copyright 2018 Optica, CC BY 4.0.)

reflectance for BSi below 6%. This is applicable to the solar industry sector, as well as in other relevant materials and device applications, such as in photodetectors and optical devices.

3.4 Summary

The above chapter presents a brief description of BSi. Recent advancements in BSi surface texturization processes are highlighted. Current applications of BSi and the modeling methodologies used to describe the optical properties of BSi that are available in the literature are discussed briefly.

References

[1] Liu X *et al* 2014 BSi: fabrication methods, properties and solar energy applications *Energy Environ. Sci.* **7** 3223–63

[2] Jansen H *et al* 1995 The BSi method: a universal method for determining the parameter setting of a fluorine-based reactive ion etcher in deep silicon trench etching with profile control *J. Micromech. Microeng.* **5** 115

[3] Soueiti J *et al* 2023 A review of cost-effective BSi fabrication techniques and applications *Nanoscale* **15** 4738–61

[4] Ayvazyan G Y 2024 Reflective properties of BSi in a wide spectral range *J. Contemp. Phys-ARME.* **59** 188–92

[5] Yoo J S *et al* 2006 BSi layer formation for application in solar cells *Sol. Energy Mater. Sol. Cells* **90** 3085–93

[6] Singh P, Sharma S N and Ravindra N 2010 Applications of porous silicon thin films in solar cells and biosensors *JOM* **62** 15–24

[7] Zhao Z *et al* 2023 BSi for near-infrared and ultraviolet photodetection: a review *APL Mater.* **11** 021107

[8] Setälä O E *et al* 2023 Boron-implanted BSi photodiode with close-to-ideal responsivity from 200 to 1000 nm *ACS Photonics* **10** 1735–41
[9] Zhao Y *et al* 2021 Light-optimized photovoltaic self-powered NO_2 gas sensing based on BSi *Sensors Actuators* B **340** 129985
[10] Song J *et al* 2023 Status and prospects of surface texturing: design, manufacturing and applications *Surf. Sci. Technol.* **1** 21
[11] Strehlke S *et al* 2000 Design of porous silicon antireflection coatings for silicon solar cells *Mater. Sci. Eng.* B **69–70** 81–6
[12] Dhoska K, Spaho E and Sinani U 2024 Fabrication of BSi antireflection coatings to enhance light harvesting in photovoltaics *Engineering* **5** 3358–80
[13] Savin H *et al* 2015 BSi solar cells with interdigitated back-contacts achieve 22.1% efficiency *Nat. Nanotechnol.* **10** 624–8
[14] Yoo J *et al* 2013 Random reactive ion etching texturing techniques for application of multicrystalline silicon solar cells *Thin Solid Films* **546** 275–8
[15] Xia Y *et al* 2011 A novel method to produce BSi for solar cells *Sol. Energy* **85** 1574–8
[16] Kim C *et al* 2015 Enhanced absorption and short circuit current density of selective emitter solar cell using double textured structure *Sol. Energy* **116** 265–71
[17] Koysuren O *et al* 2024 $SiO_2/WO_3/ZnO$ based self-cleaning coatings for solar cells *J. Sol-Gel Sci. Technol.* **110** 183–203
[18] Sreejith K P *et al* 2022 Etching methods for texturing industrial multi-crystalline silicon wafers: a comprehensive review *Sol. Energy Mater. Sol. Cells* **238** 111531
[19] Zhao J H *et al* 2020 Ultrafast laser-induced BSi, from micro-nanostructuring, infrared absorption mechanism, to high performance detecting devices *Mater. Today Nano* **11** 100078
[20] Fan Z *et al* 2021 Recent progress of BSi: from fabrications to applications *Nanomaterials* **11** 41
[21] Chapin D M, Fuller C S and Pearson G L 1954 A new silicon p–n junction photocell for converting solar radiation into electrical power *J. Appl. Phys.* **25** 676
[22] Bindra H S, Ramchandran J and Nayak R 2020 Simplified and cost-effective technique to enhance optical properties of microstructured silicon *Mater. Today Proc.* **28** 272–7
[23] Huo C, Fu H and Peng K-Q 2024 Inverted pyramid structures fabricated on monocrystalline silicon surface with a NaOH solution *Heliyon* **10** e23871
[24] Khairul Azri A A *et al* 2025 Development of surface-texturized BSi through metal-assisted chemical etching and its application in the thermophotovoltaic field: a review and recommendation *Semicond. Sci. Technol.* **40** 013001
[25] Yusuf M *et al* 2021 Optimized deep reactive-ion etching of nanostructured BSi for high-contrast optical alignment marks *ACS Appl. Nano Mater.* **4** 7047–61
[26] Miakonkikh A and Kuzmenko V 2024 Formation of BSi in a process of plasma etching with passivation in a SF_6/O_2 gas mixture *Nanomaterials* **14** 945
[27] Yang Y M *et al* 2015 Development of high-performance multicrystalline silicon for photovoltaic industry *Prog. Photovolt. Res. Appl.* **23** 340–51
[28] Vorobyev A Y and Guo C 2011 Direct creation of BSi using femtosecond laser pulses *Appl. Surf. Sci.* **257** 7291–4
[29] Pribyl M *et al* 2022 Dense high aspect ratio nanostructures for cell chip applications—fabrication, replication, and cell interactions *Micro Nano Eng.* **15** 100121
[30] Alsolami A *et al* 2024 Recent advances in BSi surface modification for enhanced light trapping in photodetectors *Appl. Sci.* **14** 9841

[31] Yao C *et al* 2024 Micro/nano-hybrid hierarchical structure of BSi decorated with gold nanoparticles for ultralow broadband reflectivity (<1%) *Appl. Surf. Sci.* **655** 159641
[32] Li J-Y, Hung C-H and Chen C-Y 2017 Hybrid BSi solar cells textured with the interplay of copper-induced galvanic displacement *Sci. Rep.* **7** 17177
[33] Vyacheslavova E A *et al* 2022 Study of cryogenic unmasked etching of 'BSi' with Ar gas additives *ACS Omega* **7** 6053–7
[34] Sanchez-Plaza G and Urquia A 2024 Process and optical modeling of BSi *Opt. Express* **32** 17704–18
[35] Rahman T and Boden S A 2017 Optical modeling of BSi for solar cells using effective index techniques *IEEE J. Photovolt.* **7** 1556–62
[36] Katsidis C C and Siapkas D I 2002 General transfer-matrix method for optical multilayer systems with coherent, partially coherent, and incoherent interference *Appl. Opt.* **41** 3978–87
[37] Pla J C *et al* 1997 Ray tracing vs electromagnetic methods in the analysis of antireflective textured surfaces [of solar cells] *Conf. Record of the 26th IEEE Photovoltaic Specialists Conf.*
[38] Ma S *et al* 2018 A theoretical study on the optical properties of BSi *AIP Adv.* **8** 035010
[39] Teixeira F L 2008 Time-domain finite-difference and finite-element methods for Maxwell equations in complex media *IEEE Trans. Antennas Propag.* **56** 2150–66
[40] Tyson J, Rahman T and Boden S 2019 Optical simulation of BSi surfaces using geometric randomisation and unit-cell based averaging *15th Photovoltaic Science, Application and Technology Conf. (April 12th 2019)* https://eprints.soton.ac.uk/432941/1/1904_Optical_Simulation_of_Black_Silicon.pdf
[41] Synopsys 2017 Sentaurus™ Interconnect User Guide Version N-2017.09, September 2017 (Mountain View, CA)
[42] Patchett S *et al* 2013 Effective index approximation for ordered silicon nanowire arrays *J. Opt. Soc. Am.* B **30** 306–13
[43] Steglich M *et al* 2014 The structural and optical properties of BSi by inductively coupled plasma reactive ion etching *J. Appl. Phys.* **116** 173503
[44] Nishijima Y *et al* 2016 Anti-reflective surfaces: cascading nano/microstructuring *APL Photonics* **1** 076104
[45] Elsayed A A *et al* 2018 Optical modeling of BSi using an effective medium/multi-layer approach *Opt. Express* **26** 13443–60

Chapter 4

Methods of processing

This chapter presents a discussion of the various techniques for processing black silicon (BSi). It addresses wet chemical etching methods, both isotropic and anisotropic, along with metal-assisted chemical etching (MACE) and electrochemical etching (ECE), highlighting recent advancements in each of the approaches. The chapter also covers dry etching techniques, including reactive ion etching (RIE), plasma immersion ion implantation (PIII), deep reactive ion etching (DRIE), and inductively coupled plasma reactive ion etching (ICP-RIE). Additionally, it discusses laser-assisted methods such as femtosecond laser etching.

4.1 Introduction

Black silicon (BSi), a modified form of crystalline silicon (Si), has enhanced light-absorbing properties, making it highly suitable for applications in photovoltaics (PVs), photodetectors, and sensors. The name 'black silicon' comes from its characteristic dark appearance, which results from nanoscale surface structures that significantly reduce the reflectivity. During recent decades, the formation of BSi and its surface has received significant attention in various research areas. Etching is one of the main techniques that is utilized to create BSi microstructures. BSi etching methods are classified based on the process mechanism and the etching medium. These methods can be broadly categorized as follows: wet etching, dry etching [1–3], and laser etching [4, 5]. Figure 4.1 shows some of the primary processing methods that are employed to process BSi.

4.2 Wet etching

Wet etching is a widely used method to create BSi due to its simplicity, cost-effectiveness, and scalability. This technique primarily involves using chemical solutions to make the silicon surface textured to reduce reflectivity and enhance light absorption. Wet etching is further classified into two types: anisotropic and isotropic etching, including electrochemical etching and metal-assisted chemical etching (MACE) (figures 4.1 and 4.2) [3]. Figure 4.2 presents a comparison of the

doi:10.1088/978-0-7503-3247-7ch4

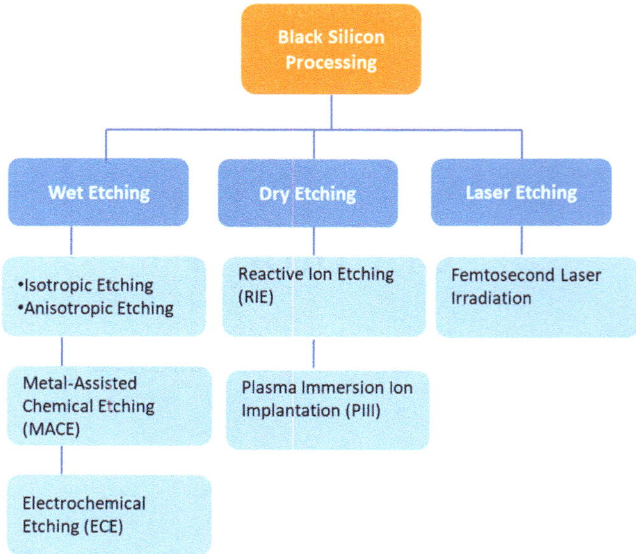

Figure 4.1. Classification of the manufacturing processes of BSi.

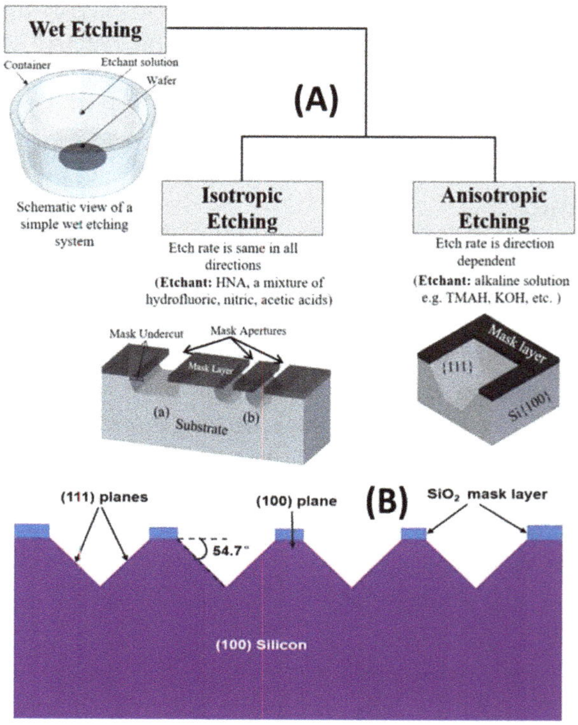

Figure 4.2. (A) Silicon etching classification used for the creation of micro/nanostructure BSi. (B) An inverted nanopyramid structure produced using anisotropic etching on a ⟨100⟩ silicon substrate. ((A) Reproduced from [3]. Copyright 2021 Springer Nature, CC BY 4.0. (B) Reproduced from [6]. Copyright 2018 IntechOpen, CC BY 3.0.)

isotropic and anisotropic etching processes. Isotropic wet etching occurs at the same rate for all crystal orientations, whereas anisotropic wet etching occurs at higher rates for certain crystal orientations [3].

4.2.1 Isotropic etching

Isotropic etching has applications in semiconductor processing for bulk material removal and cleaning silicon surfaces. It is also useful in micro-electromechanical system (MEMS) fabrication for creating smooth, curved surfaces and etching sacrificial layers for cavity formation. Isotropic etching is a process in which silicon is etched at the same rate in all directions, resulting in features with rounded or undercut profiles beneath the mask, as shown in figure 4.2(A) (a),(b) [3]. This process is straightforward and does not require advanced equipment. It is a cost-effective and scalable method for producing BSi by immersing silicon wafers in chemical solutions that create micro- or nanostructures on the surface (figure 4.2(A)). This non-directional etching is primarily used when precise geometric control is not required. The direction of etching is uniform in all directions (the x-, y-, and z-axes) and is not influenced by the crystallographic orientation of the silicon. It produces rounded or curved edges, or hemispherical shapes. It uses inexpensive chemicals compared to other methods such as plasma etching. It is suitable for industrial-scale production. Wet chemical etching relies on controlled chemical reactions between the silicon surface and an etching solution. A common etching solution is a combination of hydrofluoric acid (HF) and nitric acid (HNO_3).

HNO_3 oxidizes silicon to form silicon dioxide (SiO_2) on the surface:

$$Si + 4HNO_3 \rightarrow SiO_2 + 4NO_2 + 2H_2O. \tag{4.1}$$

HF dissolves the silicon dioxide, exposing the silicon for further oxidation:

$$SiO_2 + 6HF \rightarrow H_2SiF_6 + 2H_2O. \tag{4.2}$$

As the etching process progresses, random pits, pores, or nanostructures form, leading to a textured surface. The combined oxidation and etching cycle results in a darkened silicon surface referred to as BSi with enhanced light-trapping properties. The process parameters such as the ratio of etching solution and etching time affects the properties of BSi. The commonly used ratios of $HF{:}HNO_3{:}H_2O = 1{:}3{:}5$ or similar. Higher HNO_3 content increases oxidation, while HF concentration controls the rate of SiO_2 removal. Longer etching times produce deeper structures but can cause over-etching and weaken the wafer. Etching is typically performed at room temperature; however, elevated temperatures accelerate the reaction and may affect the uniformity of the structure. Before implementing the process of formation of BSi, silicon wafers are cleaned to remove native oxides and contaminants (e.g. using a piranha solution or RCA cleaning) to ensure uniform etching. The etching process can achieve reflectivity of BSi below 5%, enhancing the efficiency of solar cells made from such materials. In order to process an optimized BSi surface, the ratio of HF and HNO_3 concentration can be tailored to achieve the desired structure size and reflectivity.

Wet chemical etching is a foundational method for BSi fabrication due to its simplicity and adaptability. However, precise control of parameters (HF and HNO_3) is essential to optimize the process of BSi formation for specific applications. Although this process has many advantages, some challenges are such that the etched surface may not have uniform nanostructures, which can limit its reproducibility. In addition, it has other disadvantages such as lack of control over the geometry and dimensions of the etched features; further, it causes significant undercutting beneath the mask, which can affect feature accuracy and is not ideal for applications that require precise, angular geometries. The by-products that are produced during the reaction such as NO_2 gas and fluorosilicic acid (H_2SiF_6) require proper disposal.

4.2.2 Anisotropic etching

Silicon wet anisotropic etching is frequently utilized in silicon bulk micromachining in MEMS technology, semiconductor processing for integrated circuits (IC), and surface texturization for solar cell applications [1–3, 7]. Anisotropic etching has high precision and geometric control due to crystallographic selectivity; also, the associated minimal undercutting (figure 4.2(A)) leads to excellent feature definition. It is useful for complex microfabrication processes. Anisotropic etching depends on the crystal plane; for example, the silicon crystal's $\langle 111 \rangle$ plane has the slowest etch rate (highly dense and stable) whereas the $\langle 100 \rangle$ and $\langle 110 \rangle$ planes have faster etch rates. The resulting structures depend on the alignment of the mask with respect to the crystal orientation. This directional etching leads to the formation of angular profiles, such as V-shaped grooves or pyramidal cavities (figure 4.2(A) and (B)). Wet anisotropic etching uses chemicals that preferentially etch silicon along certain crystal planes due to differences in the atomic density and chemical bond strength. Figure 4.3(a) depicts the crystal plane structure of $\langle 111 \rangle$, which has one dangling bond and three back bonds. These three back bonds are strong enough to stabilize the $\langle 111 \rangle$ plane surface. This crystallographic lattice surface structure means that the Si $\langle 111 \rangle$ plane has a higher activation energy than the Si $\langle 100 \rangle$ plane and, therefore, the alkaline etchant cannot etch along the $\langle 111 \rangle$ plane. Hence, as shown in figure 4.3(b), the etching rate of the alkaline etchant solution on the Si $\langle 111 \rangle$ planar surface is effectively zero, resulting in the formation of micro-pyramidal patterns [8]. Figure 4.3(c) shows a cross-section field emission scanning electron microscopy (FESEM) image of the micro-pyramidal structures made using wet chemical anisotropic etching [9].

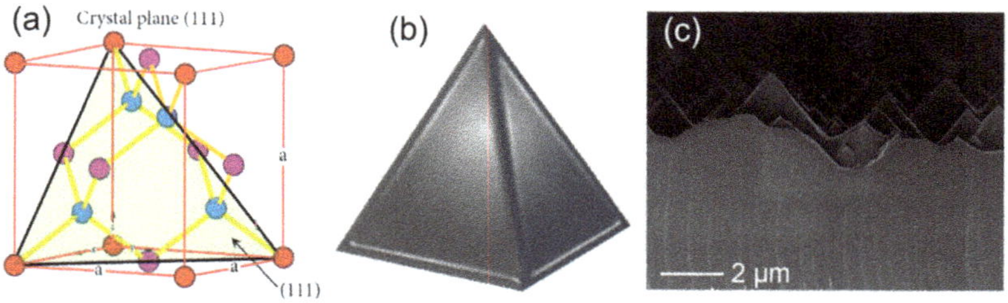

Figure 4.3. (a) Crystal plane $\langle 111 \rangle$, (b) 3D schematic of an upright pyramid, and (c) FESEM cross-sectional image of the etched sample. (Reproduced from [9]. Copyright 2021 MDPI, CC BY 4.0.)

Table 4.1. Summary of isotropic and anisotropic etching characteristics.

Feature	Isotropic etching [3]	Anisotropic etching [1–3, 7]
Etching direction	Uniform in all directions	Selective along crystal planes
Etching profile	Rounded/undercut	Angular and well-defined
Control	Less geometrical control	High precision and control
Applications	General material removal	Microfabrication and MEMS structures
Etchants used	HF/HNO_3, plasma	KOH, TMAH, EDP

The most widely used anisotropic etchant is potassium hydroxide (KOH) which has slower etch rate along the ⟨111⟩ plane compared to the ⟨100⟩ or ⟨110⟩ planes and the etching stops at the ⟨111⟩ plane, producing well-defined, angular structures. Tetra-methyl ammonium hydroxide (TMAH) has similar behavior to KOH but offers better compatibility with metals and CMOS processes; it is also less toxic and easier to handle compared to KOH. Ethylenediamine pyrocatechol (EDP) is another anisotropic etchant with a strong selectivity for crystal planes and is often used in specialized MEMS applications but is more toxic. The mask orientation determines the shape of the etched cavity such as for a mask aligned to the ⟨100⟩ orientation, a square mask results in a pyramidal cavity with ⟨111⟩ sidewalls whereas a rectangular mask can produce trapezoidal grooves. For example, for silicon wafers, etch depth and feature shape are controlled by the mask dimensions, etching solution, temperature, and etching duration. Using KOH, a typical etch angle between the ⟨111⟩ and ⟨100⟩ planes is 54.7°. Figure 4.2(B) shows that anisotropic etchants create a 54.7° angle between the ⟨100⟩ and ⟨111⟩ planes on the ⟨100⟩-oriented Si substrate [6].

Although anisotropic etching offers several benefits, it has some disadvantages such as slower etching rates along certain planes which increases processing time. Furthermore, it requires precise alignment of the mask with respect to the crystal orientation and is limited to semiconductors such as silicon and materials with strong anisotropic etching properties. A summary of some isotropic and anisotropic etching related characteristics is presented in table 4.1.

4.2.3 Metal assisted chemical etching

MACE is a technique that is used to etch silicon by leveraging a metal catalyst to enhance the etching process in a chemical solution. It is a cost-effective process and is suitable for large-scale production; it is also compatible with various orientations of silicon and doping levels [10, 11]. MACE generally includes electroless metallization of the surface, or deposition of pre-made metal nanoparticles, followed by etching with a solution containing an oxidant and a complexing agent for silicon. This simple approach has been demonstrated to effectively produce a variety of Si nanostructures, including Si nanowires (SiNWs), porous SiNWs, Si nanopores, and BSi, or high-aspect-ratio trenches with well-controlled morphological characteristics [12]. MACE overcomes the disadvantages of dry etching techniques that either need expensive instruments and consume high amounts of energy, or require complicated fabricating processes, making them

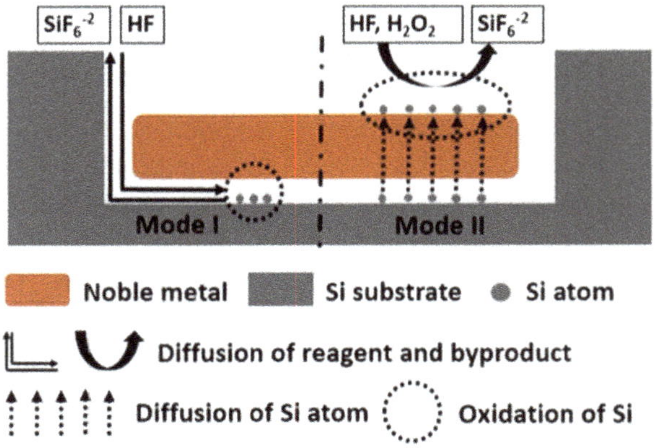

Figure 4.4. Illustration of MACE. H_2O_2 and HF serve as reactants, with the metal layer as a catalyst. The reduction of H_2O_2 generates holes, which then oxidize the silicon layer. (Reproduced from [13]. Copyright 2022 the Authors. Published by American Chemical Society. CC BY-NC-ND 4.0.)

unfavorable for industrial applications. In MACE, noble metals, for example, gold (Au), silver (Ag), platinum (Pt) or palladium (Pd), act as a catalyst, accelerating the localized oxidation and dissolution of silicon. The metal guides the etching process, creating directional etching beneath the metal pattern.

It is important to note that in conventional etching, the substrate area with the pattern (i.e. the etch mask) is protected from etching, while the area exposed to the etching solution is etched [13]. In contrast, MACE demonstrates an unusual occurrence in which the area protected by the metal is etched. This is because the metal layer acts as a catalyst, reducing oxidation while also supplying holes to the underlying Si substrate [14]. Figure 4.4 illustrates the principle of metal-assisted chemical etching. The etching process involves two main steps: metal deposition and silicon etching. First, a thin metal layer is formed on the Si substrate, typically using methods such as sputtering, evaporation, electroless plating or photolithography. Typically, noble metals such as Ag [14, 15], Au [16], Pd [17, 18], and Pt [18] are used as catalysts. The patterned substrate is then immersed in a solution of an etchant (HF) and an oxidizing agent (H_2O_2) [13]. MACE produces both isotropic and anisotropic structures, depending on the processing conditions and the crystallographic orientation of the silicon. By adjusting the process parameters, the process can control the etching depth and shape.

The metal layer on the silicon substrate works as both a catalyst and a cathode. On the metal catalyst, hydrogen peroxide (H_2O_2) is reduced to produce holes:

$$H_2O_2 + 2H^+ \rightarrow 2H_2O + 2h^+. \tag{4.3}$$

These holes diffuse through the metal layer and enter the silicon substrate, which serves as an anode. The injected holes oxidize Si, which is then dissolved at the Si–metal interface by hydrofluoric acid (HF) in the following reaction:

$$Si + 4h^+ + 6HF \rightarrow SiF_6^{2-} + 6H^+. \tag{4.4}$$

It should be noted that reaction (4.4) takes place at the interface of the Si substrate and the metal layer on top of it. In-plane, mass transfer is the most effective approach for a relatively thick metal layer (>50 nm). That is, the reaction products should diffuse along the extremely narrow area between the metal and the substrate before leaving through the edge of the metal layer (mode I in figure 4.4). This is why the MACE approach has been used primarily to create nanostructures, in which the in-plane mass transfer is significantly easier. However, in 'microstructures', the in-plane mass transfer is so slow that nonuniform etching occurs along the metal layer [14]. Because the 'in-plane' mass transfer is too slow to etch micrometer-sized metal sheets, it is proposed that enhancing 'out-of-plane' mass transfer (mode II in figure 4.4) could be a solution. When the metal layer is incredibly thin (<10 nm), pinholes in metal layer encourage out-of-plane mass transfer of reactants and products across the metal-covered substrate [13].

Uddin *et al* [19] employed the aluminum-assisted chemical etching (AACE) procedure to produce BSi. The substrate was 280 μm p-type (100) CZ single crystal silicon (1–10 Ω cm), with a thickness of 12 nm aluminum (Al) film. Figure 4.5 depicts top (figure 4.5(a)) and cross-sectional (figure 4.5(b)) view FESEM images of BSi manufactured using the AACE technique. Annealing at the optimal temperature of 400 °C resulted in BSi nanopores with the deepest (average depth of 256 nm) and largest (49%) surface coverage. The reflectance of the BSi decreased to 9.8% in the 300–1100 nm region, while $R_{\min}$ was reduced to 6.6%, as shown in figure 4.5(c) [20].

Despite its several advantages, MACE offers the following challenges: it causes metal contamination and residual metal may need to be removed post-etching to avoid contamination. Furthermore, it requires precise control over metal deposition and etching conditions. Variability in the metal catalyst and layer thickness can lead to uneven etching. Concentrations of HF and H_2O_2 control the etching rate and surface smoothness. Temperature may affect the reaction kinetics, with higher temperatures generally increasing the etch rate.

MACE is a powerful technique for fabricating silicon-based nanostructures and microstructures with applications in cutting-edge technologies.

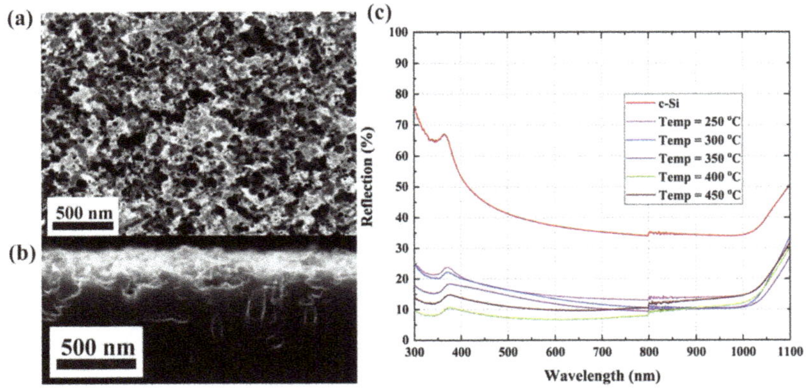

Figure 4.5. (a) Top and (b) cross-sectional views of FESEM images of BSi manufactured using the AACE technique at 400 °C annealing temperature. (c) Reflection profiles of BSi manufactured at different annealing temperatures (250 °C–450 °C) in the wavelength range of 300–1100 nm, compared to c-Si. (Reproduced with permission from [19]. Copyright 2021 Elsevier.)

4.2.4 Electrochemical etching

Electrochemical etching (ECE) of silicon is a process in which silicon is etched in an electrolyte solution using an electric current. This etching has been popularized mainly after the discovery of luminescence from porous silicon [21, 22], which currently belongs to the materials with a wide range of industrial applications [23, 24]. ECE is commonly used to create porous silicon, BSi, microstructures, or features on silicon surfaces for applications in microelectronics, photonics, sensors, and biomedical devices [25–28].

Figure 4.6 [29, 30] provides a schematic of the electrochemical etching set-up to produce BSi, in which the sample is usually held horizontally with a solid back contact, a Cu-plate, and the front side is in contact with a reservoir containing the electrolyte [29, 30]. In the electrolyte, there will be a Pt-cathode. In the set-up, the Si sample is positively biased (anode) and is in contact with the electrolyte in which a negatively biased Pt-electrode (cathode) is placed. The mechanism of pore initiation is still under debate in the literature; however, there are suggestions that defects or slight variations in surface potential due to defects or doping atoms are the starting point of the pores. When a bias is turned on, holes from the sample and fluoride (F^-) ions in the electrolyte will move towards the electrolyte–substrate interface and react. The exact reaction kinetics is not well understood, and may vary quite significantly depending on the formation parameters (i.e. current density and time) as is evident from the different morphologies of BSi. The main reaction during porous structure or BSi formation, assuming a hydrogen terminated Si surface, is suggested by Lehmann and Gosele [31] and can be written as

$$SiH_2 + 2F^- + 2h^+ \rightarrow SiF_2 + H_2 \tag{4.5}$$

$$SiF_2 + 4HF \rightarrow 2h^+ + Si_2F_6^{2-} + H_2. \tag{4.6}$$

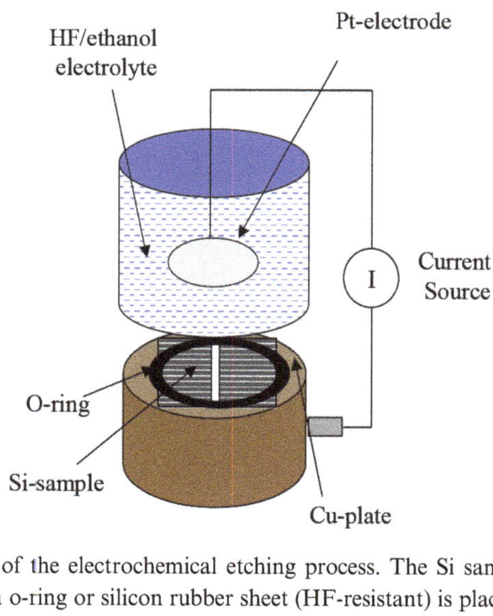

Figure 4.6. Basic schematic of the electrochemical etching process. The Si sample is placed on a Cu-plate, which is positively biased, an o-ring or silicon rubber sheet (HF-resistant) is placed on top of the sample. The electrolyte is filled into the reservoir and a Pt-electrode is placed in the electrolyte. The Pt-electrode is negatively biased. A current source controls the current/bias. (Reproduced with permission from [29].)

In reactions (4.5) and (4.6), SiH_2 is bound to the Si surface. The hydrogen gas that is formed may interfere with the etching process. In this general reaction, two holes are required for each Si atom dissolved, hence the valence of the reaction is two. After etching, the electrolyte is removed from the etch bath and the sample in it is rinsed out with ethanol. The sample is then taken out to air dry. In the electrolyte and immediately after drying, the pore walls are mostly H-terminated [32]. The hydrogen is replaced by oxygen and a native oxide is formed quite rapidly in air. This will change the properties of the BSi over time. The electrolyte contents used for BSi etching may vary substantially; however, the electrolyte is generally based on aqueous HF. In addition, to facilitate the extraction of hydrogen bubbles formed during etching, ethanol (CH_3CH_2OH) is usually used as a surfactant.

Etching parameters such as HF concentration control the rate of etching and porosity. Sometimes, H_2O is also added to the electrolyte to control both the aqueous concentration of HF and the rate of the oxidation reaction. Ethanol improves wettability by enabling the release of HF throughout the reaction as well as its penetration into silicon pores; it also reduces the surface tension.

The etching mechanism can operate in three regions based on current density: low, high, and transitional current density [33, 34]. In the low current density zone, porous silicon is formed because the etching reaction is limited by SiO_2. The high current density region creates a large number of holes across the entire region/surface of the wafer, and the reaction is restricted by the elimination of the oxide, resulting in the wafer becoming electrochemically polished. A transition region exists between the other two regions, allowing randomly distributed pillars to be produced without the use of a mask. The material parameters such as doping type, doping concentration in the Si wafer, and illumination conditions contribute to the reliance of the exact current density levels in the three regions.

In the case of BSi, the low current density zone is generally considered to be important. In this region, etching is primarily controlled by adjusting the etching duration, light, and current density. Illumination is used by n-type silicon to generate holes. The resulting silicon pore diameters mostly vary with current density. Their morphology (e.g. macroporous or nanoporous) varies according to doping type (p or n) and resistivity. A silicon wafer retains its porosity, and pores expand when a steady current is applied [35]. Furthermore, variations in current density can lead to the creation of structures with different refractive indices and porosities. The resulting structure lowers light reflection by 5% [36].

The process of etching is highly precise and can control the feature size. It is relatively simple and is a cost-effective set-up. However, few challenges include the use of HF, which is highly toxic and corrosive. Precise control of etching parameters is required for reproducibility; it is also not very compatible with large-scale manufacturing.

4.3 Dry etching

Dry etching of silicon refers to etching techniques that use gases or plasma to remove material from the silicon surface [37]. It is widely used in microfabrication due to its precision, anisotropy, and compatibility with advanced lithography techniques [38]. Dry etching can further be categorized as reactive ion etching (RIE), plasma immersion ion implantation method (PIII), deep reactive ion etching (DRIE), and inductively coupled plasma reactive ion etching (ICP-RIE) (figure 4.1).

4.3.1 Reactive ion etching

RIE is a versatile dry etching technique with several variations tailored to specific applications or materials. The next section discusses the variations of RIE, which are all plasma-based and include DRIE, PIII, ICP-RIE, and cryogenic reactive ion etching (Cryo-RIE).

The RIE system is a plasma-based tool that is used to etch materials with high precision by combining chemical reactions with physical ion bombardment. Grass-like BSi is typically formed by self-masking during RIE. The process involves the simultaneous etching and deposition of passivation layers, leading to the formation of nanoscale spikes or pits on the silicon surface. The commonly used etching gases are sulfur hexafluoride (SF_6) for etching and oxygen (O_2) for passivation. In plasma conditions, high radio frequency (RF) power, low pressure, and specific gas ratios are used.

Figure 4.7 depicts an RIE system. It consists of a vacuum chamber which is a sealed chamber in which the etching process occurs under controlled vacuum conditions. The system is equipped with ports for gas flow, RF power, and wafer loading. It has plasma source which generates plasma by applying RF energy to the gas mixture in the chamber. The gas delivery system supplies etching and passivation gases (e.g. SF_6, CF_4, O_2). The vacuum pump, which maintains the low pressure (~1–100 mTorr) required for plasma generation, is usually a combination of mechanical and turbo pumps. In-line mass-flow controllers (not visible in figure 4.7) regulate the

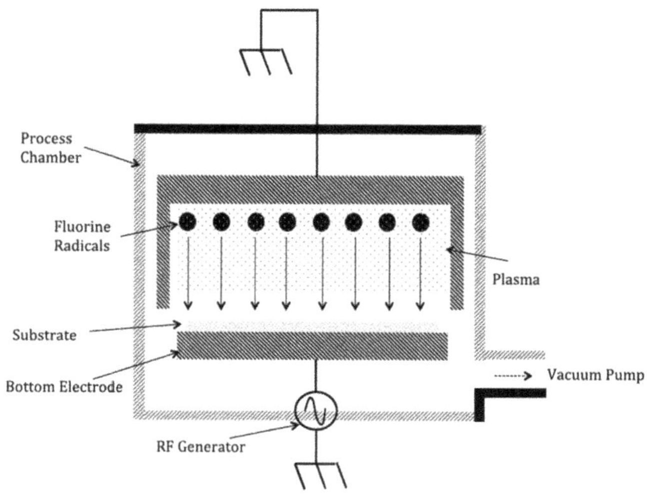

Figure 4.7. Representation of an RIE system. The inlet gas lines are not shown. (Reproduced with permission from [40]. Copyright 2020 Springer Nature.)

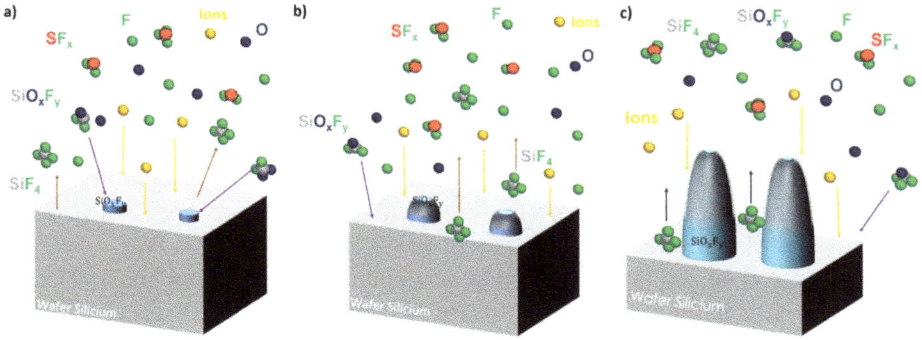

Figure 4.8. Schematics of the general principles of the physical and chemical mechanisms involved in etching, as well as their progression with time. (a) Random SiO_xF_y spot formation on the surface of the wafer. (b) Side passivation with a layer of SiO_xF_y and silicon etching with fluor radicals. (c) Nanostructures formed at the completion of the etching procedure. (Reproduced with permission from [41]. Copyright 2020 Elsevier.)

flow of process gases into the etching chamber. RIE etchers are typically single-wafer etching tools, which provide better uniformity [39]. Most commercial etchers feature load-lock chambers to increase wafer throughput. The electrodes are parallel plates in which plasma is confined; the wafer is placed on the cathode (powered electrode). This electrode is connected to an RF generator operating at frequencies of 13.56 MHz and up to ionize the gas. The chamber walls and the top electrode are electrically grounded. The electromagnetic field applied to the process gases entering the etch chamber causes electrons to be removed from the gas molecules, thus beginning the formation of plasma.

Halogens (group VII elements) such as fluorine, chlorine, bromine, and iodine are used in RIE etching operations due to their extremely electronegative and reactive nature. These halogens react with the material being etched; they produce chemical compounds known as halides. The various process gases used in RIE of various material types [39, 40] are, for example, fluorine-based gases such as SF_6 and CF_4 for silicon and silicon carbide etching.

The three mechanisms that occur during plasma etching are as follows: chemical etching, physical etching, and passivation. Figure 4.8 depicts a schematic of these three silicon etching methods [41]. This method employs SF_6 and O_2 gases to generate F* and O* radicals [42]. F* is responsible for etching silicon, producing volatile products such as SiF_x. In a competitive reaction, SiF_x products react with O* radicals (reaction (4.7)) to generate a passivation layer of SiO_xF_y that protects the surface from etching [43]. This passivation layer is partly removed by ion bombardment and the exposed silicon is further etched by F*:

$$SiF_{x(\text{ads})} + O^* \rightarrow SiO_xF_y. \tag{4.7}$$

Thus, at the start of the etching process, plasma inhomogeneity causes the development of random SiO_xF_y patches that protect the silicon from etching (figure 4.8(a)). These areas will then become the top layer of the nanostructures (figure 4.8(b)) by protecting them from etching. Furthermore, physical etching is caused by the ion bombardment on the bottom electrode during the etching process (figure 4.8). Physical etching enhances chemical etching [44]; therefore, the etching rate will differ between the side of the structures (unaffected by ion bombardment) and the silicon wafer. Because of the difference in etching rate, the sides of the structures will be protected by a thin layer of SiO_xF_y (figure 4.8(b). Finally, this etching/passivation competition mechanism causes the BSi to consist of random silicon nanostructures [45]. At the end of the procedure, nanostructures of a few hundreds of nanometers are formed, with the SiO_xF_y layer protecting the top and sides of the structures (figure 4.8(c).

RIE has several advantages, including good depth uniformity, good mask selectivity, less chemical waste handling issues (compared to wet etching), a relatively clean process, the ability to provide high fidelity and dimensional control of the etched features, and the ability to be automated for cassette-to-cassette high-wafer throughput production. Because of these desirable characteristics, among others, RIE is a commonly used etch method in the IC fabrication process [40, 46, 47]. However, some of the challenges such as achieving consistent uniform structures over large areas is difficult. The RIE process can be expensive and time-consuming for large-scale production.

4.3.2 Plasma immersion ion implantation

PIII is often used to enhance the properties of BSi by modifying its surface characteristics. PIII is indeed a variation of RIE in the sense that both are plasma-based technologies but they are designed for fundamentally different purposes. PIII is used for surface modification and ion implantation whereas RIE is used for material removal in microfabrication to create precise patterns on substrates. PIII, like RIE, uses SF_6/O_2 gas combinations which are bled into the vacuum chamber at a consistent flow rate. PIII transforms SF_6 and O_2 into reactive ions and radicals [48]. Under an applied negative voltage, reactive ions are introduced into the silicon substrate and react with it. PIII introduces dopants such as phosphorus, boron, or sulfur into the BSi to form p–n junctions or to adjust the doping profile for photovoltaic or electronic applications [49, 50]. PIII can also be used to implant elements such as nitrogen or fluorine to functionalize the surface for applications in catalysis or chemical sensors [51]. Furthermore, it is also useful in reducing surface recombination by implanting elements such as hydrogen or fluorine to create a passivated layer that minimizes defects and enhances the efficiency of BSi in photovoltaic devices. PIII modifies the surface hardness and wear resistance without compromising the light-absorbing properties of BSi. PIII can be utilized to implant silver, titanium, or other biocompatible ions to make BSi suitable for medical devices or antimicrobial applications [52]. The plasma immersion technique ensures uniform ion implantation over the textured BSi surface. It is scalable and compatible with industrial processes, making it feasible for large-scale production. In addition, it is suitable for temperature-sensitive applications, as it does not require the high temperatures that are associated with traditional doping methods.

By combining BSi's intrinsic properties with the versatility of PIII, this technique unlocks advanced functionalities for cutting-edge applications in renewable energy, biomedicine, and beyond.

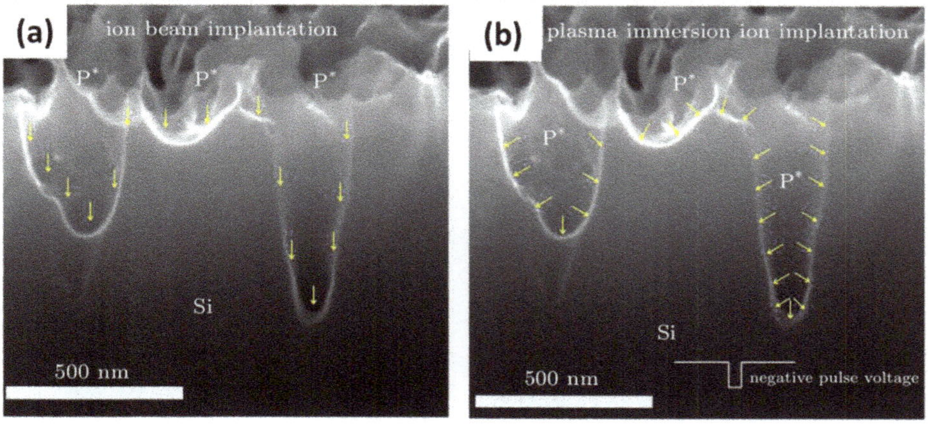

Figure 4.9. SEM image of a BSi surface with phosphorus (P) implantation models: (a) beam-line ion implanter and (b) PIII. PIII might accomplish conformal implantation on BSi. (Reproduced with permission from [53]. Copyright 2014 Chinese Physical Society and IOP Publishing.)

According to the study in [53], PIII can achieve conformal doping on mc-Si and BSi with implantation non-uniformity of less than 5%, making it acceptable for industrial mass production. The secondary ion mass spectrometer profile shows that the PIII approach results in a 100 nm shallow emitter. The study investigates the effect of PIII doping doses and annealing conditions on the performance of multicrystalline and BSi solar cells. A schematic diagram (figure 4.9) is presented to demonstrate the differences between the implantation process in beam-line implantation (figure 4.9(a)) and the plasma immersion ion implanter (figure 4.9 (b)) [53]. It is evident from figure 4.9(a) that the ions emitted from the beam-line implanter are vertical to the wafer surface but tilted at an angle to the surface of the BSi layer. However, the ions generated by PIII, as shown in figure 4.9, surround the surface morphology and are vertical to the BSi layer. With each pulse voltage, the ions are conformally implanted into the silicon [53].

4.3.3 Deep reactive ion etching

DRIE is specially designed for creating deep trenches or high-aspect-ratio structures [40]. Typically, it employs the Bosch™ process, which alternates between etching and passivation steps. DRIE was adopted by the MEMS industry as the preferred bulk silicon micromachining technology [40]. In DRIE, SF_6 and C_4F_8 are injected into the reaction chamber for etching and passivation, respectively [54]. This technique can be repeated hundreds or thousands of times to create a very deep and nearly vertical silicon needle type structure. The etched holes have almost vertical sidewalls, and the etches can penetrate the silicon substrate to depths of hundreds of microns or deeper [33, 40, 55, 56].

Figure 4.10 shows the Bosch™ technique, a version of the DRIE process [56]. The etch is a cyclical dry plasma etch method that alternates between a high-density

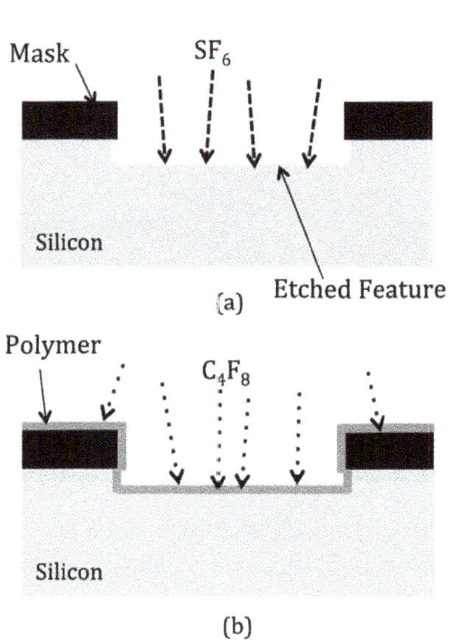

Figure 4.10. A diagram of the Bosch™ technique for DRIE etching of silicon. (a) SF_6 is utilized to produce the fluorine-based reactive species required to etch the silicon. (b) The etch tool turns off the SF_6 process gas and activates the C_4F_8 process gas. (c) The process gas C_4F_8 is turned off, and the etching gas SF_6 is turned back on. (d) SF_6 is turned off, and C_4F_8 is turned back on. (Reproduced with permission from [40]. Copyright 2020 Springer Nature.)

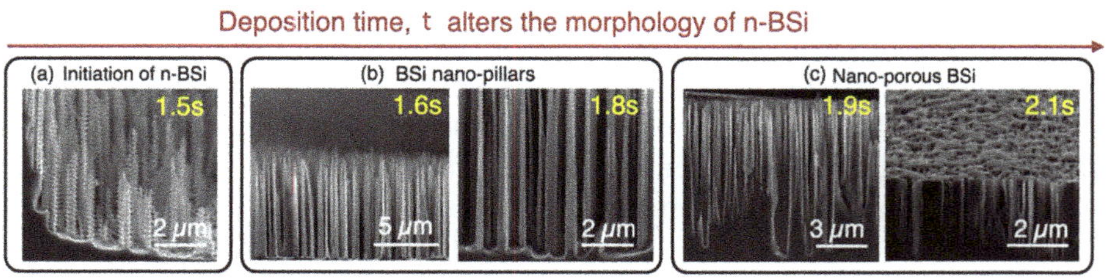

Figure 4.11. Cross-sectional SEM images show (a) initiation of nanostructured BSi at $t = 1.5$ s, (b) BSi nanopillars at t = 1.6 and 1.8 s, and (c) nanoporous BSi at $t = 1.9$ s and 2.1 s Note: the number of cycles for each deposition time was kept constant at 100. (Reproduced with permission from [57]. Copyright 2021 American Chemical Society.)

plasma to etch silicon in one cycle and depositing an etch-resistant polymer layer on the sidewalls in the next. The silicon is etched using a SF_6 chemistry, while the deposition of an etch-resistant polymer layer on the sidewalls uses a C_4F_8 chemistry.

In a related study [57], nanostructured BSi was fabricated via an optimized DRIE process. The study reveals the effect of the polymer deposition time, t, on the nanostructured BSi morphology. As shown in figure 4.11(a), at $t = 1.5$ s, nanostructured BSi formation was initiated along the etched sidewalls. BSi nanopillars were found at $t = 1.6$ and 1.8 s (figure 4.11(b)) [57]. Furthermore, nanoporous BSi is formed at $t = 1.9$ s and $t = 2.1$ s. In terms of morphology, they had pillar-like bases and needle-like tops with the typical DRIE scalloped sides. Previous investigations shed light on these findings [58].

4.3.4 Inductively coupled plasma reactive ion etching

ICP-RIE is commonly used to fabricate BSi. ICP-RIE employs two RF power sources; one for high-density plasma generation (inductive coil) and another for substrate bias to control the ion energy for anisotropic etching. The etching process provides independent control of plasma density and ion energy. The process relies on the high-density plasma generated in an ICP system, which allows for precise control over the etching process to produce the characteristic BSi morphology. It uses etching gases such as sulfur hexafluoride (SF_6) for silicon removal and oxygen (O_2) for passivation. The passivation layer is significantly influenced by silicon temperatures [59, 60]. Furthermore, Dussart's work [59] investigates BSi etching at cryogenic temperatures. Microscopic contaminants or silicon etch residues act as local masks during the process, causing uneven etching and the formation of nanoscale structures (e.g. spikes or pores). ICP causes less damage to substrates due to low-energy ion bombardment.

In a study, for the first time in the literature, BSi was optimized as a nanostructure initiator mass spectrometry (NIMS) surface using a traditional ICP etching technique with SF_6 and O_2 gas mixture [61]. The impact of substrate temperature on BSi formation is investigated by raising the temperature from −120 °C to −80 °C. At −80 °C and a SF_6/O_2 gas flow rate of 34/16 sccm/sccm, BSi is formed only after etching for more than 6.5 min. SEM images of BSi substrates produced at −120 °C and −80 °C are shown in figures 4.12(A) and (B) [61]. At higher substrate temperatures (−80 °C), SiO_xF_y passivation decreases, resulting in slower BSi formation rates [59]. Increasing the oxygen concentration compensates for this impact by increasing passivation, as illustrated in figures 4.12(A) and 4.12(C), where oxygen is increased from 16 to 20 sccm while the overall gas flow rate remains constant at 50 sccm [61]. At higher temperatures and oxygen concentrations, silicon pillars seem to be more homogeneous and thinner, with a height of ~1 μm and a high aspect ratio of ~8.7 (figure 4.12(C)), compared to low temperatures and low oxygen concentrations (figure 4.12(A)) [61]. This results in a higher surface area and, as proposed previously, lower laser intensity is required for gas phase ion generation.

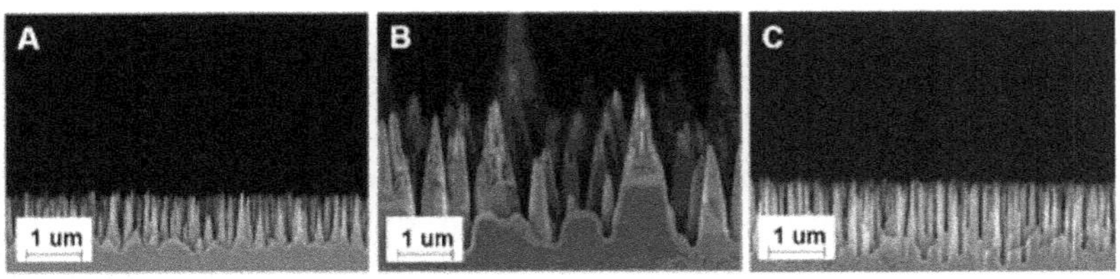

Figure 4.12. SEM images showing cross-sectional views of BSi surfaces taken under various conditions: (A) SF_6/O_2 34/16 sccm/sccm at −120 °C for 5 min; (B) SF_6/O_2 34/16 sccm/sccm at −80 °C for 6.5 min; (C) SF_6/O_2 30/20 sccm/sccm at −80 °C for 6.5 min. (Reproduced with permission from [61]. Copyright 2016 American Chemical Society.)

Another variation of RIE is cryo-RIE which involves etching at cryogenic temperatures (below −100 °C) to suppress sidewall etching [59]. Gas chemistries (e.g. SF_6 and O_2) interact with the cold substrate to create passivated sidewalls. This prevents undercutting for high-aspect-ratio structures and produces anisotropic layers with smooth sidewalls; it also has low thermal stress on the substrate.

4.4 Laser-assisted methods

In the 1990s, experiments involving laser-assisted etching of silicon surfaces led to the initial discovery of BSi. The technique was developed at Harvard University by Eric Mazur and his team. In this process, a high-powered femtosecond laser was used to etch the surface of crystalline silicon in the presence of a halogen-containing gas, sulfur hexafluoride (SF_6). The laser pulses created microscopic structures, such as sharp cones or ridges, on the silicon surface. This increased the surface area and altered its optical properties. The surface exhibited a much darker appearance, almost black (thus referred as BSi), due to its enhanced light absorption across a wide range of wavelengths. The earliest laser-induced cones in silicon were first observed and studied by Rothenberg and Kelly [62].

In laser processing, a lens focuses an ultrafast laser pulse beam (with a duration of femtoseconds (fs), picoseconds (ps), or nanoseconds (ns) onto a Si substrate, which is generally loaded in a vacuum chamber filled with a background gas [62, 63]. In order to create a large ablation region, the vacuum chamber is set to a 3D stage in advance, so that the position of the laser spot on the Si substrate may be modified and controlled by a computer program. After being irradiated with ultrafast laser pulses, periodically ordered micro-nanostructures can form on the Si surface. The creation of micro-nanostructures on the surface and the doping mechanism of BSi can be illustrated using pulsed laser–matter interactions. The interaction of ultrafast laser pulses with Si material can cause damage to the Si surface, which is controlled by the laser fluence [62, 63].

Figure 4.13 depicts the schematic diagram of the femtosecond laser processing system employed in a recent study [5]. The laser beam was focused through a lens (75 mm focal length) to irradiate a 40 μm laser spot on the silicon wafer [5]. The experimental details are available in [5]. The laser-inducing and laser-cleaning processes were carried out alternately. The detailed procedure was as follows: first, the laser scans in a specific direction to form one row of laser-induced BSi; second, adjacent scanning by laser cleaning with the same row spacing is carried out; third, the laser-induced process is repeated along the same path as the second step; and finally, large-scale uniform BSi is formed by repeating the aforementioned steps. Figures 4.13(B) and (C) illustrate the differences between this method and the typical laser induction method [5].

Mei *et al* [64] used femtosecond laser irradiation and HF etching to develop a new form of BSi with flexible characteristics. Figure 4.14(a) shows a scanning electron microscope image of BSi under femtosecond laser irradiation. Viorobyev *et al* [65] created BSi by shining a high-intensity femtosecond laser directly onto the silicon surface in an air atmosphere. Figure 4.14(b) depicts an SEM image of BSi under

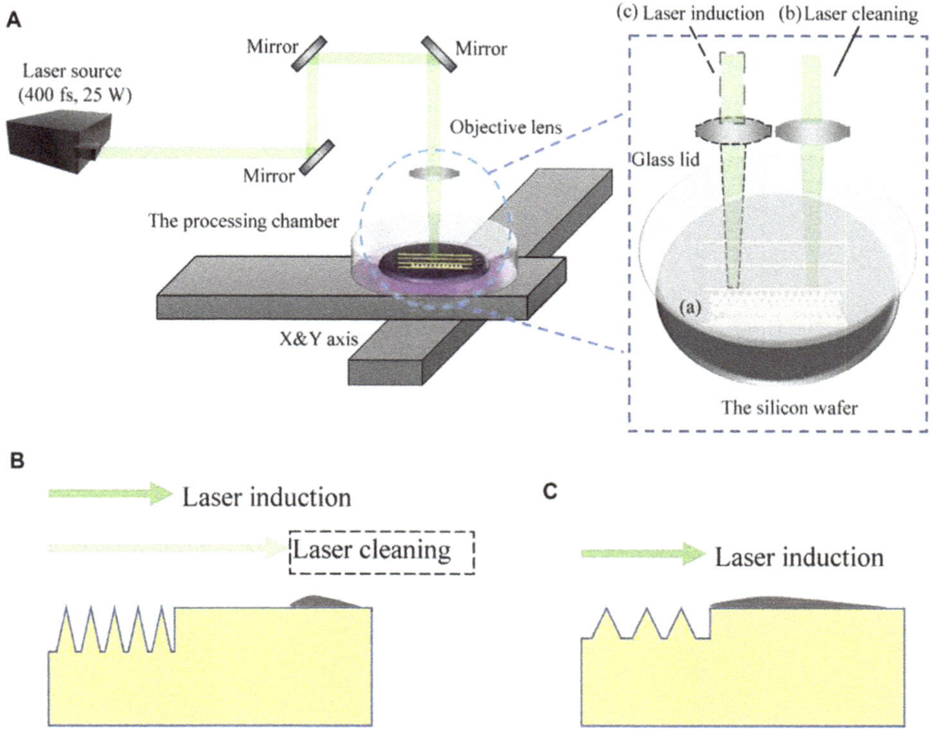

Figure 4.13. (A) Schematic illustration of the femtosecond laser processing system. (B) Laser induction combined with laser cleaning and (C) conventional laser induction. (Reproduced from [5]. Copyright 2022 Frontiers Media, CC BY 4.0.)

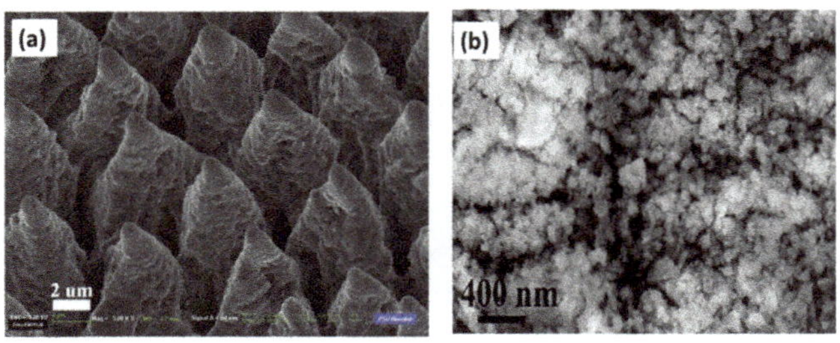

Figure 4.14. Scanning electron microscope (SEM) images of (a) the micro spike structure of the fabricated flexible BSi. (b) Nanoscale structures of treated BSi under femtosecond laser irradiation in an air atmosphere. ((a) Reproduced with permission from [64]. Copyright 2011 Elsevier. (b) Reproduced with permission from [65]. Copyright 2011 Elsevier.)

femtosecond laser irradiation in an air atmosphere. The structural region appears velvet black from all angles.

Table 4.2 provides a comparative study focusing on critical variables such as morphology, control, scalability, cost, speed, advantages, disadvantages, and applications of various aforementioned BSi techniques.

Table 4.2. Comparative analysis of BSi surface modification techniques; metal-assisted chemical etching (MACE), reactive ion etching (RIE), electrochemical etching (ECE), deep reactive ion etching (DRIE), and laser-assisted methods such as femtosecond laser etching.

Technique	Morphology	Control	Scalability	Cost	Speed	Advantages	Disadvantages	Applications
MACE [10–15, 19]	Nanowires, nanostructures	Moderate	High	Low	Moderate	Simple, low-cost, high aspect ratio, large-area processing	Nonuniform structures, uses hazardous chemicals (e.g. HF)	Photovoltaics, photodetectors, sensors
RIE [39–41, 46, 47]	Nanopillars, pores	High	Moderate	High	Moderate	Precise control, uniform features, very low reflectance	Expensive equipment, requires optimization	Solar cells, antireflective surfaces, optics
ECE [21, 22, 25–28, 36]	Porous silicon	High	Moderate	Moderate	Slow	High porosity control, simple set-up, moderate cost	Limited to small substrates, requires doped silicon	Biosensors, porous membranes, drug delivery
DRIE [33, 40, 55–57]	Deep trenches, pillars	Very high	Low	Very high	Slow	Deep and anisotropic etching, highly uniform structures	High equipment cost, limited scalability	MEMS, microfluidics, precision optics
Laser-assisted etching [5, 62, 63, 65]	Micro/ nanospikes, cones	Very high	Very high	High	Fast	No chemicals, precise and customizable patterns, suitable for large areas	Expensive equipment, risk of thermal damage	Large-area coatings, sensors, photonic devices

4.5 Summary

This chapter explored BSi processing techniques, including wet chemical etching, dry etching, and femtosecond laser-assisted methods, emphasizing recent advancements. Wet chemical etching methods include the following: isotropic, anisotropic, metal-assisted chemical etching (MACE), and electrochemical etching (ECE). Reactive ion etching (RIE), plasma immersion ion implantation (PIII), deep reactive ion etching (DRIE), and inductively coupled plasma reactive ion etching (ICP-RIE) are examples of dry etching methods for the formation of BSi.

References

[1] Gupta A, Pal P and Sharma C S 2019 Surface texturing of silicon {100} in an extremely low concentration TMAH for minimized reflectivity *ECS J. Solid State Sci. Technol.* **8** P622

[2] Pal P and Sato K 2017 *Silicon Wet Bulk Micromachining for MEMS* (Singapore: Jenny Stanford)

[3] Pal P *et al* 2021 High speed silicon wet anisotropic etching for applications in bulk micromachining: a review *Micro Nano Syst. Lett.* **9** 4

[4] Liu X *et al* 2022 Perspectives on black silicon in semiconductor manufacturing: experimental comparison of plasma etching, MACE, and fs-laser etching *IEEE Trans. Semicond. Manuf.* **35** 504–10

[5] Wen Z *et al* 2022 Large-scale black silicon induced by femtosecond laser assisted with laser cleaning *Front. Phys.* **10** 862605

[6] Amalathas A P and Alkaisi M M 2018 Fabrication and replication of periodic nanopyramid structures by laser interference lithography and UV nanoimprint lithography for solar cells applications *Micro/Nanolithography* ed J Thirumalai (Rijeka: IntechOpen) ch 2

[7] Solanki C S 2015 *Solar Photovoltaics: Fundamentals, Technologies and Applications* (Delhi: PHI Learning)

[8] Chen W *et al* 2018 Difference in anisotropic etching characteristics of alkaline and copper based acid solutions for single-crystalline Si *Sci. Rep.* **8** 3408

[9] Arafat M Y *et al* 2021 Study of black silicon wafer through wet chemical etching for parametric optimization in enhancing solar cell performance by PC1D numerical simulation *Crystals* **11** 881

[10] Peng K Q *et al* 2006 Fabrication of single-crystalline silicon nanowires by scratching a silicon surface with catalytic metal particles *Adv. Funct. Mater.* **16** 387–94

[11] Peng K *et al* 2006 Metal-particle-induced, highly localized site-specific etching of Si and formation of single-crystalline Si nanowires in aqueous fluoride solution *Chem. Eur.* J. **12** 7942–7

[12] Han H, Huang Z and Lee W 2014 Metal-assisted chemical etching of silicon and nanotechnology applications *Nano Today* **9** 271–304

[13] Nur'aini A and Oh I 2022 Deep etching of silicon based on metal-assisted chemical etching *ACS Omega* **7** 16665–9

[14] Geyer N *et al* 2012 Model for the mass transport during metal-assisted chemical etching with contiguous metal films as catalysts *J. Phys. Chem.* C **116** 13446–51

[15] Li L *et al* 2014 Uniform vertical trench etching on silicon with high aspect ratio by metal-assisted chemical etching using nanoporous catalysts *ACS Appl. Mater. Interfaces* **6** 575–84

[16] Song Y *et al* 2014 In-plane and out-of-plane mass transport during metal-assisted chemical etching of GaAs *J. Mater. Chem.* A **2** 11017–21

[17] Volovlikova O *et al* 2019 Investigation of the Pd nanoparticles-assisted chemical etching of silicon for ethanol solution electrooxidation *Micromachines* **10** 872

[18] Li X and Bohn P W 2000 Metal-assisted chemical etching in HF/H_2O_2 produces porous silicon *Appl. Phys. Lett.* **77** 2572–4

[19] Uddin S, Hashim M R and Pakhuruddin M Z 2021 Effects of annealing temperature towards properties of black silicon fabricated by aluminium-assisted chemical etching *Mater. Sci. Semicond. Process.* **133** 105932

[20] Zhao Z *et al* 2023 Black silicon for near-infrared and ultraviolet photodetection: a review *APL Mater.* **11** 021107

[21] Fauchet P M 1996 Photoluminescence and electroluminescence from porous silicon *J. Lumin.* **70** 294–309

[22] Cullis A G, Canham L T and Calcott P D J 1997 The structural and luminescence properties of porous silicon *J. Appl. Phys.* **82** 909–65

[23] Králik M, Goraus M and Pinčík E 2020 Optical properties of electrochemically etched N-type silicon wafers for solar cell applications *J. Electr. Eng.* **71** 406–12

[24] Föll H *et al* 2002 Formation and application of porous silicon *Mater. Sci. Eng.* R **39** 93–141

[25] Santos A and Kumeria T 2015 Electrochemical etching methods for producing porous silicon *Electrochemically Engineered Nanoporous Materials: Methods, Properties and Applications* ed D Losic and A Santos (Cham: Springer International) pp 1–36

[26] Barillaro G, Nannini A and Piotto M 2002 Electrochemical etching in HF solution for silicon micromachining *Sens. Actuators* A **102** 195–201

[27] De La Luz Merino S *et al* 2013 Applications of porous silicon formed by electrochemical etching using an electrolyte based on HF:formaldehyde *Appl. Phys.* A **111** 1077–83

[28] Saverina E A *et al* 2020 Porous silicon preparation by electrochemical etching in ionic liquids *ACS Sustain. Chem. Eng* **8** 10259–64

[29] Singh P 2009 Fabrication, characterization and other related studies for performance improvement of crystalline silicon solar cells *PhD thesis* Jamia Millia Islamia & National Physical Laboratory, New Delhi

[30] Singh P, Sharma S N and Ravindra N 2010 Applications of porous silicon thin films in solar cells and biosensors *JOM* **62** 15–24

[31] Lehmann V and Gösele U 1991 Porous silicon formation: a quantum wire effect *Appl. Phys. Lett.* **58** 856–8

[32] Smith R L and Collins S D 1992 Porous silicon formation mechanisms *J. Appl. Phys.* **71** R1–R22

[33] Hsu C-H *et al* 2014 Fabrication and characteristics of black silicon for solar cell applications: an overview *Mater. Sci. Semicond. Process.* **25** 2–17

[34] Ma L L *et al* 2006 Wide-band 'black silicon' based on porous silicon *Appl. Phys. Lett.* **88** 171907

[35] Yoo J S *et al* 2006 Black silicon layer formation for application in solar cells *Sol. Energy Mater. Sol. Cells* **90** 3085–93

[36] Otto M *et al* 2015 Black silicon photovoltaics *Adv. Opt. Mater.* **3** 147–64

[37] Miakonkikh A and Kuzmenko V 2024 Formation of black silicon in a process of plasma etching with passivation in a SF_6/O_2 gas mixture *Nanomaterials* **14** 945

[38] Plakhotnyuk M M *et al* 2017 Low surface damage dry etched black silicon *J. Appl. Phys.* **122** 143101
[39] Huff M 2021 Recent advances in reactive ion etching and applications of high-aspect-ratio microfabrication *Micromachines* **12** 991
[40] Huff M 2020 *Process Variations in Microsystems Manufacturing* (Berlin: Springer)
[41] Atteia F *et al* 2020 Morphologies and optical properties of black silicon by room temperature reactive ion etching *Mater. Res. Bull.* **131** 110973
[42] Kumaravelu G, Alkaisi M M and Bittar A 2002 Surface texturing for silicon solar cells using reactive ion etching technique *Conf. Record of the 29th IEEE Photovoltaic Specialists Conf., 2002* pp 258–61
[43] Jansen H *et al* 1996 A survey on the reactive ion etching of silicon in microtechnology *J. Micromech. Microeng.* **6** 14
[44] Coburn J W and Winters H F 1979 Ion- and electron-assisted gas-surface chemistry—an important effect in plasma etching *J. Appl. Phys.* **50** 3189–96
[45] Jansen H V *et al* 2010 Black silicon method XI: oxygen pulses in SF_6 plasma *J. Micromech. Microeng.* **20** 075027
[46] Wolf S 1995 *Silicon Processing for the VLSI Era* (Sunset Beach, CA: Lattice)
[47] Campbell S A 2001 *The Science and Engineering of Microelectronic Fabrication* (New York: Oxford University Press)
[48] Möller H J *et al* 2005 Multicrystalline silicon for solar cells *Thin Solid Films* **487** 179–87
[49] Young D L *et al* 2016 Low-cost plasma immersion ion implantation doping for Interdigitated back passivated contact (IBPC) solar cells *Sol. Energy Mater. Sol. Cells* **158** 68–76
[50] Vervisch V *et al* 2006 Plasmaimmersion ion implantation applied to P+N junction solar cells *AIP Conf. Proc.* **866** 253–6
[51] Cisternas M *et al* 2020 Study of nitrogen implantation in Ti surface using plasma immersion ion implantation and deposition technique as biocompatible substrate for artificial membranes *Mater. Sci. Eng.* C **113** 111002
[52] Anders A 2000 *Handbook of Plasma Immersion Ion Implantation and Deposition* **vol 8** (New York: Wiley)
[53] Ze-Nan S, Yang X, Bang-Wu L, Jin-Hu L, Chao-Bo L and Yong-Tao L 2014 Realization of conformal doping on multicrystalline silicon solar cells and black silicon solar cells by plasma immersion ion implantation *Chin. Phys.* B **23** 118801
[54] Liu X *et al* 2014 Black silicon: fabrication methods, properties and solar energy applications *Energy Environ. Sci.* **7** 3223–63
[55] McAuley S A *et al* 2001 Silicon micromachining using a high-density plasma source *J. Phys.* D: Appl. Phys. **34** 2769
[56] Laermer F and Schilp A 1996 Method of anisotropically etching silicon *US Patent* 5,501,893, Issued on March 26, 1996
[57] Yusuf M *et al* 2021 Optimized deep reactive-ion etching of nanostructured black silicon for high-contrast optical alignment marks *ACS Appl. Nano Mater.* **4** 7047–61
[58] Leopold S *et al* 2011 Formation of silicon grass: nanomasking by carbon clusters in cyclic deep reactive ion etching *J. Vac. Sci. Technol.* B **29** 011002
[59] Dussart R *et al* 2004 Passivation mechanisms in cryogenic SF_6/O_2 etching process *J. Micromech. Microeng.* **14** 190

[60] Nguyen K N *et al* 2013 On the optical and morphological properties of microstructured black silicon obtained by cryogenic-enhanced plasma reactive ion etching *J. Appl. Phys.* **113** 194903

[61] Gao J *et al* 2016 Application of black silicon for nanostructure-initiator mass spectrometry *Anal. Chem.* **88** 1625–30

[62] Sarnet T *et al* 2008 Femtosecond laser for black silicon and photovoltaic cells *Proc. SPIE* **6881** 688119

[63] Zhao J H *et al* 2020 Ultrafast laser-induced black silicon, from micro-nanostructuring, infrared absorption mechanism, to high performance detecting devices *Mater. Today Nano* **11** 100078

[64] Mei H *et al* 2011 Development of novel flexible black silicon *Opt. Commun.* **284** 1072–5

[65] Vorobyev A Y and Guo C 2011 Direct creation of black silicon using femtosecond laser pulses *Appl. Surf. Sci.* **257** 7291–4

Chapter 5

Thermal properties of black silicon

This chapter provides a comprehensive analysis of the thermal properties of black silicon (BSi), including heat capacity, thermal conductivity, and phonon scattering, with relevant examples. It also explores infrared absorption, heat dissipation characteristics, and temperature-dependent photoluminescence of BSi. Additionally, various applications leveraging its thermal properties are discussed, such as thermoelectric, photo-thermoelectric, and photovoltaic technologies. Emerging and future applications, including thermophotovoltaics, thermal sensing, infrared detectors, passive radiative cooling, and aerospace applications, are also discussed briefly.

5.1 Introduction

Black silicon (BSi) is a modified form of crystalline silicon (c-Si) with a nanostructured surface that enhances its optical, electrical, and thermal properties. These properties make BSi valuable for applications in photovoltaics, sensors, and thermal management [1–4]. Understanding the thermal properties of BSi is critical for optimizing its performance in photovoltaic (PV), infrared (IR) sensing, thermoelectric (TE), photo-thermoelectric (PTE), and thermophotovoltaic (TPV) devices. As described in chapter 4, BSi can be fabricated by a variety of techniques such as reactive ion etching (RIE), laser processing, or metal-assisted chemical etching (MACE), that lead to the ultra-low reflectance and high absorption of BSi across a wide spectral range [2, 5, 6]. Its high absorptance leads to greater heat accumulation compared to conventional silicon, affecting temperature-dependent efficiency in solar cells. Furthermore, the nanostructure of BSi influences its thermal conductivity, primarily by increasing phonon scattering, which reduces heat dissipation efficiency [7, 8]. These properties necessitate careful thermal management strategies to optimize the performance of BSi in energy and optoelectronic applications. In addition, understanding the thermal behavior of BSi is essential for its integration into high-efficiency PV systems and other emerging technologies that require controlled heat transfer. Ongoing research aims to enhance its thermal stability

doi:10.1088/978-0-7503-3247-7ch5

while maintaining superior optical and electrical performance, ensuring its viability for the next-generation energy and sensing solutions.

5.2 Fundamentals

In general, a thorough understanding of the thermal properties is important for selecting materials that effectively manage heat transfer. This section outlines the fundamental equations for evaluating the thermal behavior of nanostructured BSi versus bulk c-Si, highlighting their importance in various applications with relevant examples.

5.2.1 Heat capacity

Heat capacity (C) measures a material's ability to store heat and is defined as the ratio of heat added or removed to the resulting temperature change, excluding phase transitions [9]:

$$C = \frac{\Delta Q}{\Delta T} = \frac{\mathrm{d}Q}{\mathrm{d}T}(\mathrm{deg}^{-1}). \tag{5.1}$$

Heat capacity is influenced by atomic vibrations, known as phonons. In solids, individual atomic vibrations are not independent but instead interact, leading to waves of atomic displacements. These waves are defined by their wavelength (λ) and frequency (ν). For a given frequency, the smallest unit of vibrational energy, known as a phonon, is given by $h\nu$. Thermal energy in a crystal is the collective energy of all phonons (or vibrational waves) at a given temperature (T). Besides atomic vibrations, thermal excitation of electrons can also influence heat capacity. However, for valence electrons to contribute significantly, they must absorb energy of the order of $\sim k_{\mathrm{B}}T$. Since only a small fraction of electrons near the Fermi level participate in the electronic transitions, their contribution to heat capacity is minimal and negligible at room temperature [10].

5.2.2 Thermal conductivity

The thermal conductivity (k), of a material defines its capability to transfer heat. It represents the rate at which heat energy moves through a material in response to a temperature gradient and is measured in watts per meter per kelvin (W m^{-1} K^{-1}). Typically, heat flows from higher to lower temperature regions within a material. This behavior is governed by Fourier's law [11], and is written mathematically as

$$q = -k\nabla T, \tag{5.2}$$

where q is the heat flux (W m^{-2}) and refers to the rate at which thermal energy passes through a unit area over a given time. k is the thermal conductivity, also known as the coefficient of thermal conductivity of the material and ∇T is the temperature gradient (K m^{-1}). The negative sign, on the right hand side in equation (5.2), indicates that heat flows from higher to lower temperatures.

Heat transfer in materials occurs through both phonons (lattice vibrations) and electrons (free charge carriers), with k being the combined result of these two mechanisms. The total thermal conductivity is given by the sum of these contributions [12, 13]:

$$k = k_L + k_E, \tag{5.3}$$

where k_L and k_E represent the lattice and electronic thermal conductivities, respectively [12, 13]. Lattice conductivity arises from heat transfer by phonons. Electronic conductivity occurs when free electrons in the conduction band, which are in thermal equilibrium with lattice vibrations in hotter regions, migrate to cooler regions. As they move, they transfer part of their thermal energy back to the lattice through phonon scattering. In insulators and semiconductors, heat is mainly transported by phonons due to the lack of free electrons [14].

5.2.3 Phonon scattering

BSi exhibits reduced thermal conductivity compared to bulk c-Si due to enhanced phonon scattering caused by its nanostructured surface. The mechanisms responsible for this reduction include phonon–boundary scattering, phonon–phonon interactions, and defect-induced scattering, all of which shorten the phonon mean free path. While bulk c-Si has a thermal conductivity of around 148 W m^{-1} K^{-1} at room temperature, the nanoscale surface modifications in BSi significantly alter its heat transport properties. The interplay of these scattering mechanisms is crucial for its applications in electronics, thermoelectrics, and thermal management.

Phonon scattering is thoroughly covered in several textbooks [8, 15, 16]. Here, a brief overview of different phonon scattering mechanisms is provided [8, 15–17].

5.2.3.1 Phonon–boundary scattering

The nanostructured morphology of BSi introduces numerous surface irregularities that serve as scattering centers for phonons, the primary heat carriers in non-metallic solids. As phonons propagate through the material and encounter surfaces, interfaces, or grain boundaries, their paths are disrupted, reducing their mean free path and consequently lowering the thermal conductivity [17]. A common approach to modeling this process assumes that the phonons are reflected upon reaching the boundary, as depicted in figure 5.1(a) [17]. The scattering rate is expressed as [18] $\tau_b{}^{-1} = b_s v_g / L$, where b_s is a geometrical factor, and L denotes the characteristic length of the structure, such as the nanowire diameter or nanofilm thickness.

This effect is particularly pronounced in structures with feature sizes that are comparable to or smaller than the phonon mean free path, as boundary scattering becomes a dominant mechanism in impeding thermal transport. Studies have shown that in silicon nanostructures, boundary scattering significantly reduces the thermal conductivity, with the extent of reduction depending on the surface roughness and structural dimensions [19].

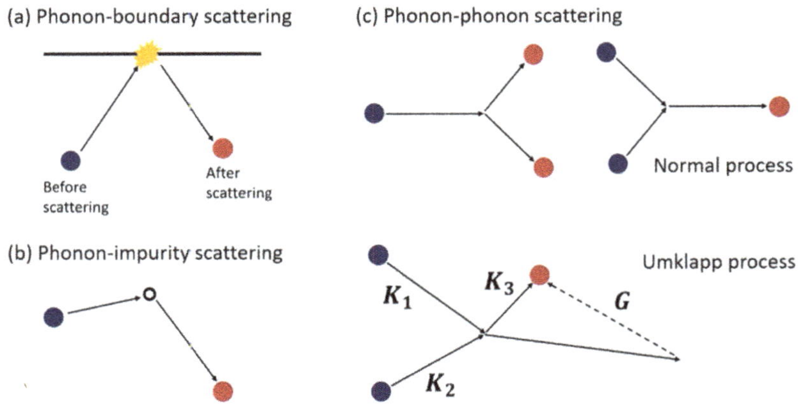

Figure 5.1. Schematic representations of various phonon scattering mechanisms: (a) phonon–boundary scattering, (b) phonon–impurity scattering, and (c) phonon–phonon scattering, including both the normal and Umklapp processes. (Reproduced with permission from [17]. Copyright 2020 Wiley.)

5.2.3.2 Phonon–impurity scattering

In the presence of impurities within the crystal structure, phonons are scattered by impurity sites such as atomic defects, vacancies, or isotopes, as illustrated in figure 5.1(b). When the defect size is significantly smaller than the phonon wavelength (i.e. the impurities act as point defects), the scattering cross-section follows the Rayleigh law, scaling as $\sigma \propto \lambda^{-4}$ or $\propto \omega^4$. Consequently, the reciprocal of the impurity scattering time is given by $\tau_\mathrm{i} = A\omega^4$ [20].

5.2.3.3 Phonon–phonon interactions

In addition to boundary scattering, intrinsic phonon–phonon interactions, especially Umklapp processes, play a crucial role in determining the thermal conductivity of BSi. Umklapp scattering involves phonon collisions that result in a net momentum change, effectively resisting heat flow. The most common type of phonon–phonon scattering involves three-phonon interactions, which include normal (N) and Umklapp (U) processes, as illustrated in figure 5.1(c). In the N process, two phonons collide to form a third phonon, or a single phonon splits into two, with both energy and momentum conserved. As a result, the N process does not directly contribute to thermal resistance. In contrast, the U process involves two phonons with wavevectors K_1 and K_2 combining to form a third phonon, whose wavevector is folded back by the reciprocal lattice vector G. While energy remains conserved, momentum is not, leading to resistance in heat flow by disrupting the movement of energy carriers [17].

Deriving an exact analytical model for the scattering rate of U processes is complex. However, an approximate expression is commonly used, given by $\tau U^{-1} = \beta\omega^n T^m$, where β, n, and m are constants [16]. The rate of these scattering events increases with temperature, leading to a further reduction in the thermal conductivity at elevated temperatures. Although often overlooked, four-phonon and higher-order phonon scattering processes [21] can significantly impact thermal transport properties, as highlighted by recent research. In four-phonon interactions, three phonons combine into one, a single phonon splits into three, or two phonons interact and transform into two new phonons. Similar to three-phonon processes, four-phonon interactions can include both normal and Umklapp processes, depending on whether phonon momentum is conserved [17, 21]. For instance, in materials such as silicon, four-phonon scattering has been shown to reduce the thermal conductivity by up to 30% at high temperatures, aligning theoretical predictions more closely with experimental observations [21].

5.2.4 Effect of nanostructuring on thermal conductivity: case studies

The most significant effect of nanostructuring is the enhancement of phonon suppression [22]. Various structures, as shown in the right column of figure 5.2, have been designed to tailor the thermal conductivity by enhancing phonon scattering [23]. The left column of figure 5.2 illustrates the different scattering mechanisms that can be targeted in nanostructured materials. The scattering rate ($1/\tau$), which is the reciprocal of effective phonon relaxation time (τ), is estimated using Matthiessen's rule [24], which combines different phonon scattering mechanisms by summing the reciprocals of their individual relaxation times [17, 23]:

$$\frac{1}{\tau_{\text{total}}} = \sum \frac{1}{\tau_i} = \frac{1}{\tau_i} + \frac{1}{\tau_b} + \frac{1}{\tau_U} + \frac{1}{\tau_e}. \tag{5.4}$$

Here 'i', 'b', 'U', and 'e' represent impurity scattering, boundary scattering, Umklapp scattering, and phonon–electron scattering, respectively. In most non-metallic materials, phonon–electron scattering has little impact on the thermal conductivity, while reduced phonon scattering enhances k, as seen in diamond and silicon [17, 23]. In metals [25, 26] and heavily doped semiconductors [27], heat is primarily conducted by free electrons, with electron–phonon interactions playing a crucial role. The Wiedemann–Franz law explains the high thermal conductivity of materials such as copper and silver by linking the electrical and thermal conductivity [28].

Heat conduction in micro- and nanostructures is complex due to phonon interactions with grain boundaries, interfaces, and confinement effects. In silicon nanostructures, hotspots smaller than the phonon mean free path cause localized heating, increasing the thermal resistance and disrupting phonon equilibrium. Bulk silicon thermal conductivity values are generally unsuitable for nanoscale applications. Doping introduces impurity scattering, reducing thermal conductivity, while grain boundary scattering further lowers it at room temperature [36–42]. In thin silicon films, thermal conductivity depends on thickness and microstructure, with phonon–boundary scattering reducing it by over 10% and up to 50% for ~100 nm films [8, 43]. Additionally, thermal energy utilization of BSi presents opportunities for various applications [44].

Several studies have reported the measurements of thermal conductivity of BSi using techniques such as time-domain thermoreflectance (TDTR) [45] and micro-Raman spectroscopy [46]. Research indicates that BSi exhibits thermal conductivity

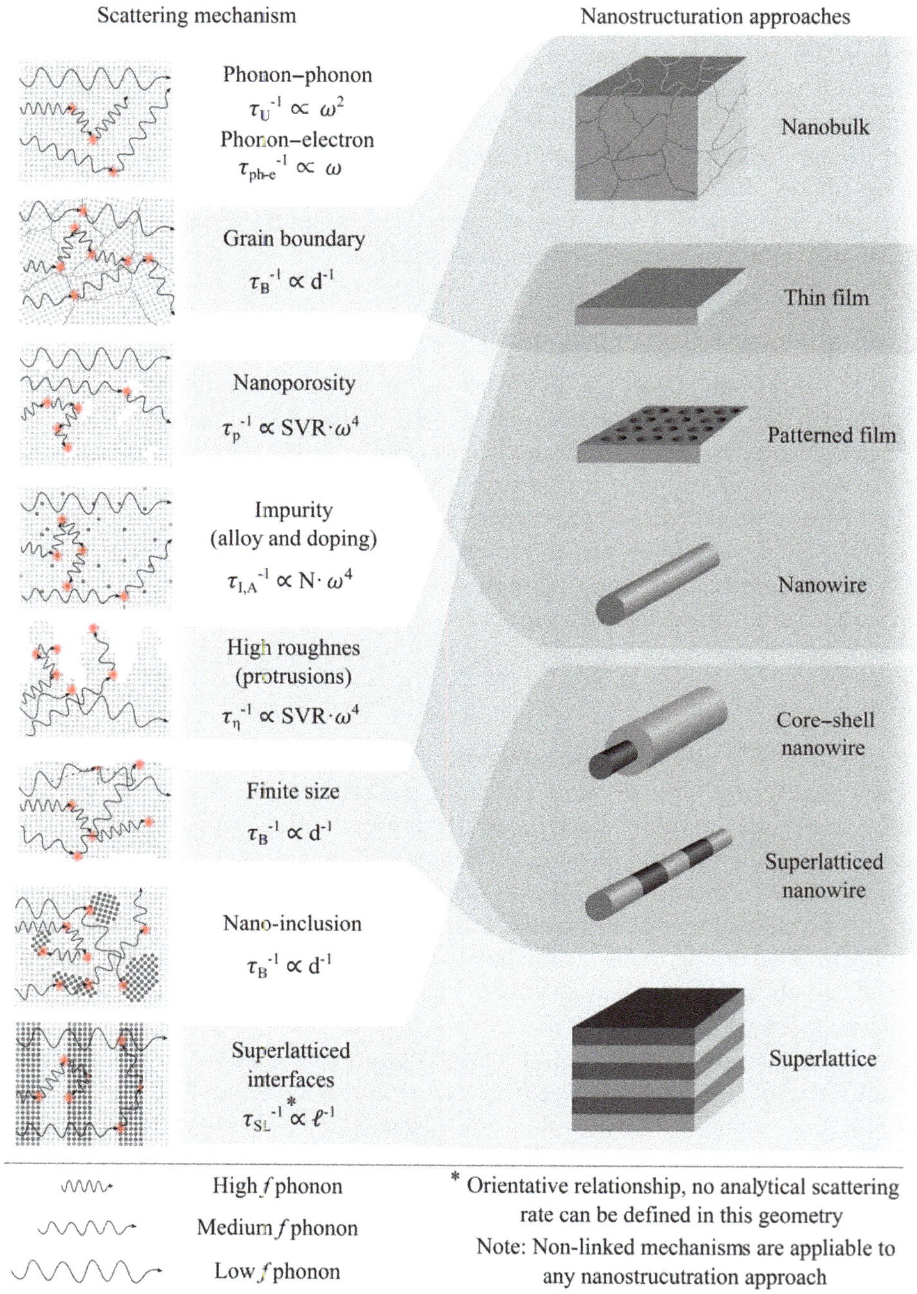

Figure 5.2. A schematic representation of different phonon scattering mechanisms, accompanied by a summary of important nanomaterial strategies for optimizing thermal transport. The links indicate the specific scattering mechanisms targeted by each nanostructure. Representative phonon scattering relationships have been derived from multiple sources [29–35]. (Reproduced from [23]. Copyright 2023 AIP Publishing, CC BY 4.0.)

values ranging from 1 to 50 W m^{-1} K^{-1}, depending on the fabrication process and feature size.

Thermal conductivity as a function of temperature is studied [8] in a research paper via a simple model based on kinetic theory:

$$k = \frac{1}{3} C_p v_s \Lambda. \tag{5.5}$$

The average phonon group velocity (v_s) is assumed to be constant at 6533 m s^{-1} [47], representing a weighted average of the transverse and acoustic phonon velocities. The phonon mean free path (Λ) represents the average distance a phonon travels between scattering events which varies significantly with temperature. According to the Debye model [48], the specific heat (C_p) follows a cubic temperature (T^3) dependence at low temperatures ($T < 0.1\theta_D$ where θ_D= 645 K for silicon). In this range, most phonons scatter at film boundaries due to a low phonon population and minimal intrinsic scattering events [49]. When the phonon mean free path is approximately equal to the film thickness, thermal conductivity scales with T^3. At temperatures above θ_D, the specific heat remains constant, and phonon–phonon Umklapp scattering becomes the primary heat transport mechanism.

Figure 5.3(a) presents the measured thermal conductivities of intrinsic single c-Si nanowires with diameters of 22, 37, 56, and 115 nm. Compared to bulk silicon, [51] two important features are measured that are common to all the nanowires [50]: (i) the thermal conductivity of nanowires is approximately two orders of magnitude lower than that of bulk silicon, and it decreases as wire diameter decreases. This indicates that enhanced boundary scattering significantly affects phonon transport in silicon nanowires. (ii) The peak thermal conductivity shifts to higher temperatures as the wire diameter decreases, occurring at approximately 210 K, 160 K, and 130 K for 37, 56, and 115 nm wires, respectively, in contrast to the 25 K peak for bulk silicon. This shift suggests that, with decreasing diameter, phonon–boundary scattering becomes dominant over phonon–phonon Umklapp scattering, leading

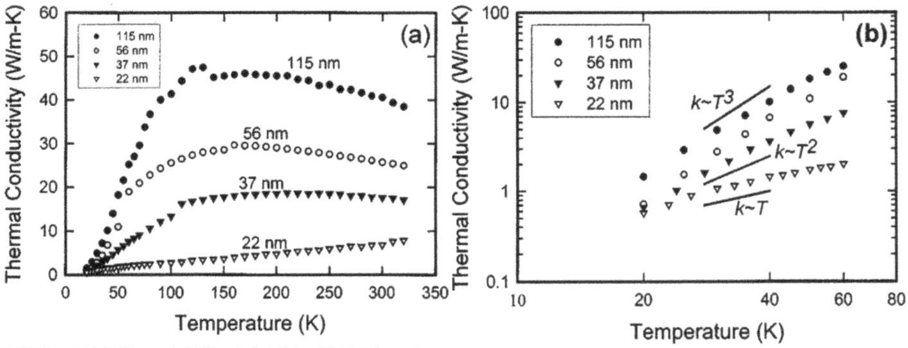

Figure 5.3. (a) Measured thermal conductivity of silicon nanowires with different diameters. The number next to each curve indicates the corresponding wire diameter. (b) Low-temperature experimental data presented on a logarithmic scale, along with T^3, T^2, and T^1 curves for comparison. (Reproduced with permission from [50]. Copyright 2003 AIP Publishing.)

to reduced thermal conductivity at higher temperatures. Notably, the 22 nm wire does not show a peak within the measured temperature range [50].

Figure 5.3(b) presents the temperature dependence of the thermal conductivity between 20 and 60 K on a logarithmic scale. It can be observed that the data for the 115 and 56 nm diameter wires align well with the Debye T^3 law in this temperature range. This indicates that boundary scattering, which is independent of both frequency and temperature, is the primary phonon scattering mechanism, causing the thermal conductivity to follow the temperature dependence of specific heat. As the nanowire diameter decreases, boundary scattering is expected to dominate further. However, for the 37 and 22 nm nanowires, the power exponent decreases with decreasing diameter, and the 22 nm wire shows a clear deviation from Debye T^3 law. This suggests that additional effects beyond phonon–boundary scattering influence thermal transport in smaller nanowires [50].

5.3 Infrared absorption and heat dissipation

The nanostructured surface of BSi enhances light absorption and IR emissivity, improving radiative heat dissipation and cooling efficiency. Its high absorption in the near-IR (NIR) and mid-IR range contributes to localized heating, requiring thermal management in applications such as solar cells. With ultra-low reflectance and high absorptance across the ultraviolet, visible, and NIR wavelengths, BSi is well-suited for PV, photodetectors, thermal sensing, energy harvesting, and TPV [52].

As discussed in chapter 7, the light trapping in BSi improves through multiple scattering, gradient refractive index, and surface plasmon effects, achieving over 95% absorptance across the solar spectrum. The nanostructures extend the optical path length, increasing photon absorption, while the refractive index gradient minimizes Fresnel reflections. Studies confirm that BSi significantly outperforms c-Si in light absorption, making it a promising material for high-efficiency optoelectronic applications [53]. The increased absorptance of BSi is primarily due to the formation of needle-like microstructures on the silicon surface. Studies have demonstrated that BSi can achieve near-unity absorption in the NIR spectrum, extending up to wavelengths of 2.5 μm. This characteristic is particularly advantageous for applications in photodetectors and IR imaging systems, where efficient IR absorption is crucial [52].

In order to address NIR absorbance attenuation in silicon, a method for depositing gold nanoparticles (AuNPs) on micro–nano-structured BSi is proposed [54]. By leveraging the local surface plasmon resonance (LSPR) of AuNPs, the study achieves BSi with broad-spectrum and high absorption.

Figure 5.4 shows the absorption curves of nanostructured BSi with 10–25 nm AuNPs. The addition of AuNPs significantly enhances absorption in the 1100–2500 nm range. BSi with 25 nm AuNPs exhibits higher absorption, averaging 98.6% between 400 and 1100 nm and 97.8% from 1100 to 2500 nm. Compared to BSi without AuNPs, absorption in the 1100–2500 nm range increases from 90.1% to 97.8%, effectively mitigating NIR absorptance attenuation. BSi shows great

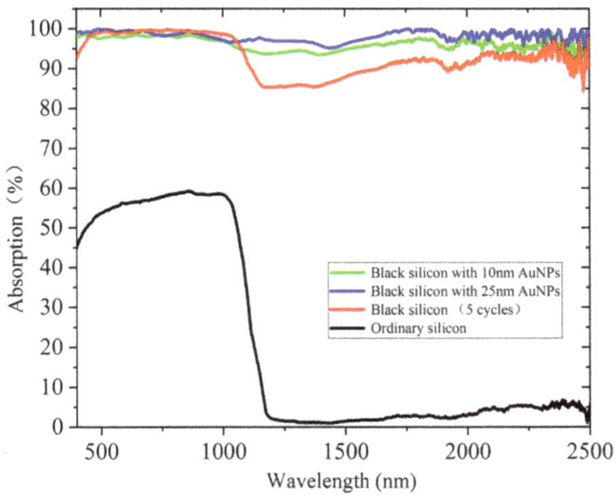

Figure 5.4. Absorption curves of BSi deposited with 10–25 nm gold nanoparticles (AuNPs) (micro–nano-structured BSi formed by five-cycle femtosecond laser ablation). (Reproduced from [54]. Copyright 2023 Springer Nature, CC BY 4.0.)

potential for NIR-enhanced photoelectric detection and micro-optical night vision imaging applications [54].

Although enhanced IR absorption of BSi improves optoelectronic device performance, heat generation is increased. The nanostructured morphology of BSi aids heat dissipation through increased surface area for efficient thermal radiation and convective heat transfer, along with high IR emissivity that enhances radiative cooling, benefiting passive cooling and thermal management in electronic devices [55].

The low thermal conductivity of BSi poses heat dissipation challenges, necessitating strategies such as thin-film integration with high-conductivity substrates, high-emissivity coatings for radiative cooling, and microfluidic cooling for active heat removal [56]. Its high IR absorption requires careful thermal management to prevent overheating, which could impact device performance and longevity. Despite these challenges, ongoing research supports its potential in infrared sensing, imaging, and thermal management applications.

5.4 Temperature-dependent photoluminescence

Photoluminescence (PL), discussed in chapter 7, in BSi exhibits notable temperature-dependent behavior, which has been systematically studied to understand its optical properties [57]. BSi, fabricated by irradiating silicon wafers with femtosecond laser pulses, displays a broad PL band centered around 630 nm (1.968 eV) at low temperatures (10 K). In the low-temperature range (10–120 K), the PL intensity decreases gradually as temperature increases. This decline is attributed to thermal quenching mechanisms, where nonradiative recombination processes become more prominent with rising temperature. In the high-temperature range (above 120 K), more pronounced quenching of PL intensity occurs. This intensified decrease is linked to increased carrier mobility, leading to enhanced nonradiative

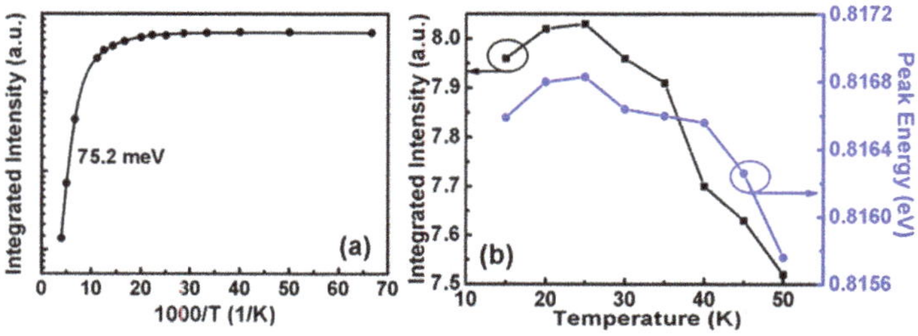

Figure 5.5. (a) Integrated PL intensity plotted against the reciprocal of temperature. (b) Variation of integrated PL intensity and peak energy with temperature in the range of 15–50 K. (Reproduced with permission from [58]. Copyright 2013 Optica Publishing Group.)

recombination and thermalization effects. At a low temperature of 10 K, the PL intensity exhibits a sublinear relationship with excitation laser intensity. This sublinearity suggests that the PL originates from recombination processes involving defect-related band tail states [57]. The visible and near-infrared PL spectrum of BSi makes it a promising material for short-haul optical communications [57].

Figure 5.5(a) presents the integrated PL intensity as a function of the reciprocal of temperature, revealing a high quenching temperature with PL detectable up to 250 K [58]. The thermal activation energy (E_a) is 75.2 meV, indicating potential applications at room temperature. Figure 5.5(b) illustrates the integrated PL intensity and peak energy as a function of temperature in the low-temperature range of 15 K–50 K. Notably, as the temperature rises from 15 K to 25 K, the PL intensity increases instead of decreasing, while the peak energy exhibits a blueshift rather than a redshift, indicating anomalous behavior [58]. This phenomenon can be attributed to the increased diffusion length of photogenerated carriers at low temperatures. As the temperature rises, carriers diffuse more efficiently, allowing a greater number to be captured by dislocation-related bound states. The accumulation of carriers leads to a filling effect, strengthening the PL intensity and causing a blueshift in peak energy. However, as the temperature continues to rise, nonradiative recombination dominates due to thermal quenching, resulting in a decrease in PL intensity, as depicted in figure 5.5(b).

These findings [57, 58] highlight the complex interplay between temperature and PL properties in BSi, emphasizing the role of defect states and nonradiative processes in determining its optical behavior.

5.5 Applications

The low thermal conductivity of BSi influences applications such as thermoelectric energy conversion, optoelectronic heat management, and infrared sensing.

5.5.1 Thermoelectric

Reduced thermal conductivity benefits thermoelectric devices by preserving temperature gradients, enhancing heat-to-electricity conversion efficiency. Nanostructured silicon surpasses conventional TE materials such as bismuth telluride in abundance, environmental sustainability, and high-temperature stability (>250 °C). However, its inherently high thermal conductivity restricts thermoelectric efficiency, resulting in a low figure of merit (Z). In order to improve its performance, a substantial reduction in the thermal conductivity is essential [59].

In a related study, silicon pillars have been explored for TE applications [59]. Silicon pillars are fabricated using a wafer-scale top-down process, involving inductively coupled plasma cryogenic dry etching, thermal oxidation, and oxide stripping. Pillars with diameters ranging from 2 μm to 170 nm and heights between 26.7 and 32.1 μm were analysed. 3ω measurements using a Wollaston probe demonstrated that reducing the pillar diameter to 700 nm led to a 27% decrease in thermal conductivity compared to bulk silicon, showing promise for improved thermoelectric applications [59].

Thermal conductivity measurements were conducted on over 100 silicon pillars using the 3ω method with scanning electron microscopy (SEM), employing a Wollaston wire probe—a 10 μm platinum wire with a 1.2 mm exposed section, stripped of its silver cladding. A thermal wave is generated by passing an alternating current at frequency ω through the Wollaston wire, penetrating the sample based on its thermal conductivity. This alters the temperature of wire and consequently its resistance. The temperature change in the wire occurs at 2ω and is measured using the 3ω voltage, which corresponds to the signal amplitude at 3ω, decreasing along the wire. The real part of the complex oscillation amplitude can be approximated based on its frequency dependence:

$$V_{3\omega} \propto \frac{1}{k}. \tag{5.6}$$

In a semi-phenomenological approach, Matthiessen's rule [24] can be applied to determine the thermal conductivity of a nanopillar as a function of its diameter (d) [60]:

$$\frac{1}{k} = \frac{3}{Cv_g\Lambda} + \frac{3}{Cv_g d}, \tag{5.7}$$

where Cv_g represents the product of heat capacity, C, and group velocity, (v_g), and Λ is the phonon mean free path (250 nm). A reported value of 1.7×10^9 W m^{-2} K^{-1} for the heat capacity–group velocity product was determined using empirical data from silicon nanowires.

The thermal conductivity values for different pillar diameters are shown as squares in figure 5.6, with the calculated curve from equation (5.7) superimposed as a line. The theoretical calculation aligns well with the measured data, showing a relative deviation of 2.7% for 1000 nm pillars and 0.4% for 700 nm pillars. Additionally, recently published thermal conductivity values of silicon nanowires are included as circles to further support the semi-phenomenological model [61].

Another example of silicon nanowires for TE application has been reported [63]. One dimensional silicon (1D-Si) nanowires exhibit lower thermal conductivity than bulk silicon, as confirmed by molecular dynamics simulations, enhancing their

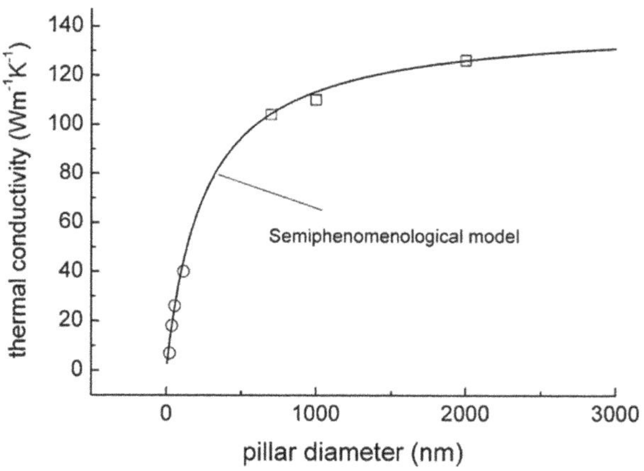

Figure 5.6. Thermal conductivity as a function of pillar diameter, calculated using the semi-phenomenological model (equation (5.7), solid line) and experimentally derived values from silicon pillars of varying diameters (open squares—Wollaston wire in contact with a single silicon pillar). For comparison, data points from silicon nanowires grown via the vapor–liquid–solid (VLS method) [62] were also included (open circles). (Reproduced with permission from [59]. Copyright 2011 Elsevier.)

potential for thermoelectric energy conversion. The calculated thermal conductivity of bulk silicon is 226.54 ± 18.17 W m^{-1} K^{-1}, which closely matches the previously reported value of 226.07 ± 2.86 W m^{-1} K^{-1} by Zhou *et al* [64]. Figure 5.7 shows that the thickness of 2D silicon nanosheets has little effect on the thermal conductivity, averaging 106 W mK^{-1}. In contrast, the thermal conductivity of a silicon nanowire (Si-NW) along the ⟨100⟩ direction is 39.62 ± 4.51 W m^{-1} K^{-1}. Previous theoretical studies reported values of 64.5 ± 4.5 W m^{-1} K^{-1} for the ⟨110⟩ [64] direction and 43 ± 10 W m^{-1} K^{-1} for the ⟨100⟩ [65] direction. The thermal conductivity of Si-NWs is considerably lower than that of silicon nanosheet (Si-NS) and bulk materials, demonstrating a significant decrease as dimensionality is reduced.

Therefore, the reduction in dimensionality is a promising approach to enhancing the ZT coefficient, a dimensionless figure of merit defined as

$$ZT = S^2 T\sigma / k, \tag{5.8}$$

where S is the Seebeck coefficient, and σ is the electrical conductivity at temperature T. As a result, silicon nanomaterials and diamond-structured nanowires synthesized under pressure show potential for advanced thermoelectric applications.

5.5.2 Photothermoelectric

Photo-thermoelectric (PTE) detectors, which rely on photothermal and thermo-electric conversion, are increasingly vital for infrared light detection. BSi, with its excellent optical properties and high absorption in the visible–NIR range, serves as an ideal light absorber for photothermal conversion in PTE detectors. In a related study [44], an NIR PTE device using BSi was developed. PTE device efficiently absorbs NIR light and converts it into thermal energy, which is then transformed

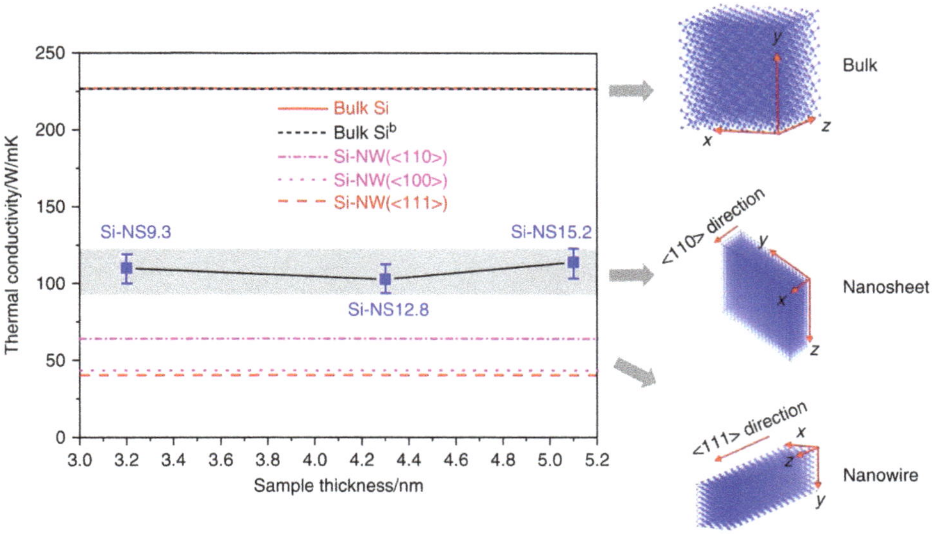

Figure 5.7. Calculated thermal conductivity of bulk silicon (Si) and Si nanostructures, including bulk Si, bulk Si^b (predicted by Zhou [64]), ⟨11⟩ direction Si nanowire, ⟨110⟩ surface Si nanowire (predicted by Zhou [64]), ⟨100⟩ direction Si nanowire (predicted by Donadio [65]), and silicon nanosheets (Si-NS) with thicknesses of 9.3, 12.8, and 15.2 nm. The gray area represents the fluctuation range of thermal conductivity for nanosheets. The insets illustrate the model systems studied in this work, including bulk silicon, a silicon nanosheet with a ⟨110⟩ zone axis orientation, and a silicon nanowire with its major dimension along the ⟨111⟩ direction. (Reproduced from [63]. Copyright 2018 Springer Nature, CC BY 4.0.)

into electricity via a thermoelectric module. Fabricated using MACE technology, BSi achieves over 90% absorption in the visible and near-infrared regions. This device exhibited approximately 50% higher output current than nonstructured silicon counterparts, demonstrating strong potential for energy conversion and infrared sensing applications [44].

Another recent study [66] proposes chimney-like textured BSi to enhance PTE detector performance. Fabricated using a two-step etching process, the BSi device achieves up to 99.5% absorption in the 220–1000 nm wavelength range. The deposition of AuNPs on BSi enhances absorption beyond 1100 nm through the localized surface plasmon resonance (LSPR) effect, reaching 34.4% at 1550 nm. BSi, with and without AuNPs, is integrated into a thermoelectric module to fabricate PTE devices whose performance is evaluated under various laser wavelengths. As shown in figure 5.8, PTE devices are represented as PD1 (without AuNPs) and PD2 (with AuNPs). The maximum output voltage of the device is 9.6 mV at 950 nm for PD1, increasing to 14.5 mV with AuNPs deposition (PD2). After Au loading, the rise time (τ_r) and decay time (τ_d) values are 25.78 and 27.22 s, respectively.

These findings demonstrate that BSi significantly enhances photothermal conversion efficiency and extends the electrical response of the device to the visible and near-infrared spectrum, offering a promising advancement in PTE technology [44, 66].

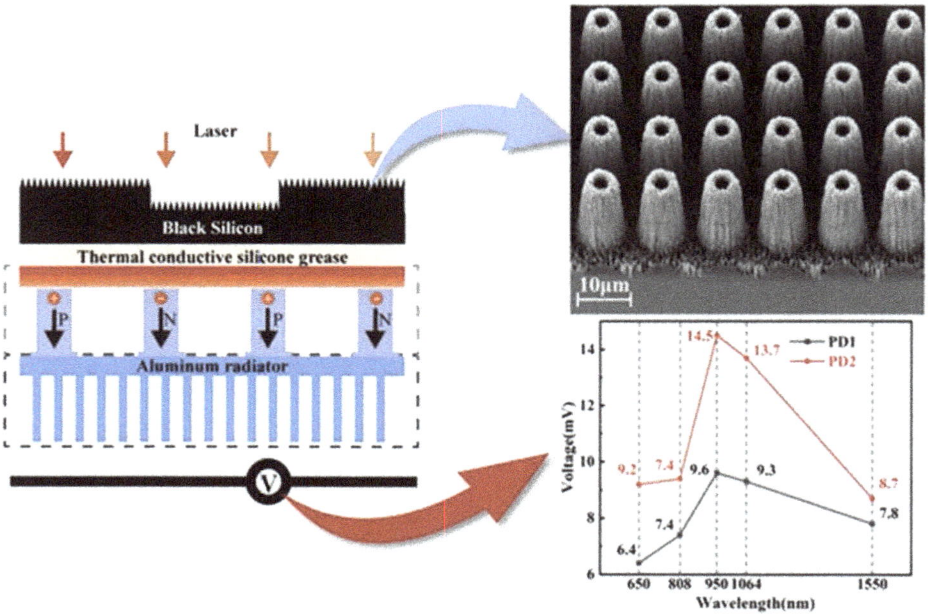

Figure 5.8. Schematic illustration of a near-infrared PTE device utilizing BSi as an absorber. The absorbed laser energy is converted into thermal energy and transferred through thermal conductive silicone grease to an aluminum radiator, where a thermoelectric module generates an output voltage. The top-right inset shows an SEM image of the nanostructured BSi surface. The bottom-right graph presents the voltage response of two photodetectors (PD1 and PD2) as a function of the incident wavelength, demonstrating enhanced voltage output at near-infrared wavelengths. (Reproduced with permission from [66]. Copyright 2024 Elsevier.)

5.5.3 Photovoltaics

One of the most promising applications of BSi is in photovoltaics, where its high absorptance enhances solar energy conversion. BSi improves solar cell efficiency by maximizing light absorption and minimizing reflection, enhancing solar energy conversion. While high absorptance improves efficiency, it also leads to increased heat accumulation. Overheating can degrade solar cell performance and reduce its lifespan. Heat generation requires effective thermal management to prevent performance degradation. The high IR emissivity of BSi enables radiative cooling, which can help lower operating temperatures and maintain efficiency over time. Its effective thermal radiation makes it a strong candidate for passive cooling technologies, reducing reliance on active cooling systems [67]. Research has shown that optimizing the surface structure of BSi can balance light absorption and heat dissipation, improving its long-term performance in solar energy applications.

The effect of annealing temperature on the surface morphology and optical properties of BSi, produced using the aluminum-assisted chemical etching (AACE) method, is investigated [7]. Annealing at 400 °C results in the deepest nanopores, highest surface coverage, and the lowest broadband reflection (300–1100 nm). This leads to the highest average absorption enhancement of 1.61 compared to the planar c-Si reference. The relationship between the average depth of nanopores and their

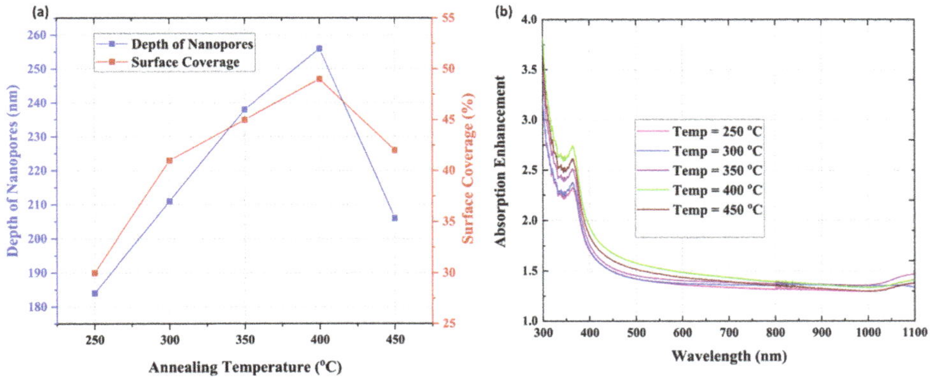

Figure 5.9. (a) Relationship between the average depth of nanopores and surface coverage of BSi against different annealing temperatures (250 °C–450 °C). (b) Absorption enhancement of BSi at different annealing temperatures (250 °C–450 °C) within the 300–1100 nm wavelength region. (Reproduced with permission from [7]. Copyright 2021 Elsevier.)

surface coverage, as a function of annealing temperature (250 °C–450 °C), is illustrated in figure 5.9(a). Both parameters increase from 250 °C to 400 °C, reaching peak values of ~255 nm depth and ~50% surface coverage at 400 °C. Beyond this, both the decreases are likely due to pore coalescence or material restructuring. The optimal annealing temperature for maximizing porosity appears to be around 400 °C. The trends observed in figure 5.9(a) provide insights into the thermal effects on nanopore characteristics, which are critical for optimizing the material properties of BSi for specific applications [7].

Figure 5.9(b) shows the absorption enhancement of BSi at different annealing temperatures, normalized to a planar c-Si reference over 300–1100 nm. The results indicate that absorption enhancement increases with higher annealing temperatures, demonstrating improved broadband light absorption. The graph shows absorption enhancement across 300–1100 nm for different annealing temperatures. Absorption is highest at short wavelengths (300–400 nm), peaking at 400 °C due to optimal nanopore formation. Beyond 500 nm, the differences between temperatures are minimal. At 450 °C, absorption declines, likely due to structural degradation. The optimal annealing temperature for maximum absorption enhancement is 400 °C [7].

In another study [68], black multi-crystalline silicon (bmc-Si) is successfully fabricated using plasma immersion ion implantation. Figure 5.10(a) presents the external quantum efficiency (EQE) of bmc-Si solar cells at diffusion temperatures of 830 °C, 825 °C, and 815 °C. As the diffusion temperature decreases, EQE improves below ~550 nm. Higher diffusion temperatures lead to increased doping levels due to higher diffusion coefficients, which in turn enhance Auger recombination. This increased recombination negatively impacts EQE at shorter wavelengths. Figure 5.10(b) shows the variation of reverse saturation current (I_o) and ideal diode factor (n) with diffusion temperature. As the diffusion temperature decreases, I_o and n also decrease, indicating reduced recombination current. However, when the temperature drops below 820 °C, both values begin to rise. Interestingly, I_o and n in

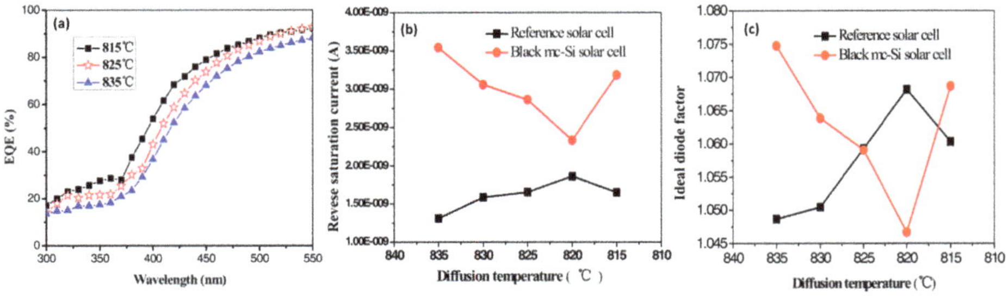

Figure 5.10. (a) The EQE of bmc-Si solar cells at different diffusion temperatures. (b) Reverse saturation current of bmc-Si and reference solar cells as a function of diffusion temperature. (c) Ideal diode factor of bmc-Si and reference solar cells at different diffusion temperatures. (Reproduced with permission from [68]. Copyright 2012 Elsevier.)

bmc-Si solar cells behave oppositely compared to reference cells, likely due to differences in surface microstructure. The optimal conversion efficiency of bmc-Si solar cells was 15.50% at a diffusion temperature of 825 °C. Notably, diffusion effects differ between bmc-Si and acid-etched mc-Si, highlighting distinct material characteristics.

5.6 Emerging innovations and future prospects

The previously mentioned applications of BSi utilize its high IR absorptance, enhanced emissivity, and modified thermal conductivity, making it highly beneficial for photovoltaics and thermal management. Its nanostructured surface reduces thermal conductivity through phonon scattering, leading to benefits in thermo-electrics and microelectronics. Moreover, its potential extends to emerging fields such as thermophotovoltaics, thermal sensing, IR detectors, passive radiative cooling, and space technology. Continued research focuses on refining its properties to drive advancements in energy conversion, sensing, and aerospace applications.

5.6.1 Thermophotovoltaics

BSi enhances the performance of TPV systems by efficiently absorbing infrared radiation and converting thermal energy into electricity. Its high absorptance improves solar cell efficiency, while strong IR emissivity aids thermal management by dissipating excess heat. These properties make it ideal for high-temperature TPV applications, with potential in industrial waste heat recovery and space power generation [60].

A recent study explores the MACE technique for BSi and its application in TPV systems [69]. Its simplicity, cost-effectiveness, and scalability position it as a promising method for TPV development. The versatility of MACE enables the fabrication of tunable BSi absorber–emitter pairs and even complete TPV cells. By optimizing the process parameters, selecting appropriate materials, and building on previous research, it could pave the way for high-efficiency TPV systems that are capable of exceeding the Shockley–Queisser efficiency limit [69].

5.6.2 Thermal sensing and infrared detectors

As discussed earlier, high IR absorption and emissivity of BSi make it ideal for thermal imaging and TPV applications [66, 70]. In addition, it is used in bolometers and thermopiles for precise temperature measurements, where efficient thermal radiation is essential. Broadband IR absorption of BSi enhances IR photodetectors for night vision and security applications. Its high emissivity enables efficient heat dissipation, preventing overheating and improving sensor stability. These properties make it ideal for high-sensitivity infrared detection in defense, medical diagnostics, and environmental monitoring.

The nanostructured surface of BSi significantly increases its sensitivity to NIR and short-wave infrared (SWIR) radiation, surpassing standard silicon detectors. This enhancement is crucial for applications requiring high sensitivity in the NIR and SWIR ranges. Its thermal transport properties are critical for thermal sensors and precise heat dissipation control.

5.6.3 Passive radiative cooling

The high-emissivity and nanostructured surface of BSi make it ideal for passive radiative cooling. It efficiently radiates heat, reducing surface temperatures, with applications in energy-efficient buildings, electronics cooling, and urban-scale cooling solutions. Its reduced thermal conductivity aids thermal insulation in microelectronics, while its high surface area enhances convective cooling. These properties support improved heat dissipation in the next-generation microprocessors, light emitting diodes (LEDs), and high-performance computing systems.

Recent work [67] investigates BSi as a potential material for nighttime passive radiative cooling, a key technology for decarbonization and energy-efficient heat management. Due to its enhanced visible and infrared absorption and emission properties, BSi significantly outperforms flat silicon in cooling performance. The results show that at 30 °C, BSi achieves a cooling power of 140 W m^{-2}, nearly 1.8 times higher than the cooling power −75 W m^{-2} of c-Si. Different BSi nanostructure depths are analysed to optimize radiative cooling efficiency. The study highlights that factors such as conductive and convective effects, atmospheric transmittance, and operating temperature significantly impact cooling performance. Notably, BSi can surpass a material with an ideal emissivity of 1 in the 8–13 μm infrared range under certain conditions [67]. Ongoing research is focused on the influence of atmospheric transmittance, with future efforts aimed at enhancing the emissivity of BSi selectivity to further optimize its cooling efficiency.

5.6.4 Space and aerospace applications

The controlled thermal properties of BSi make it ideal for utilization in space technology, helping maintain thermal equilibrium in satellites and instruments by efficiently absorbing and emitting IR radiation. Its high absorptance enhances space-based solar panel efficiency, maximizing energy conversion. In aerospace applications, BSi can improve the longevity and reliability of spacecraft that are exposed to intense solar radiation.

It is used in coronagraph instruments to suppress stray light for better celestial observations [71] and as an infrared absorber for thermal management in space instruments [72]. Its micro–nanostructures enable nearly 100% absorption in the visible and near-infrared spectrum, enhancing the performance of space-based optical systems [6].

Future research in BSi aims to enhance performance through passive cooling strategies, tailored nanostructures, and large-scale integration. As demand grows in the energy, electronics, and aerospace sectors, the global BSi technology market is projected to reach $8.9 billion by 2032, growing at a compound annual growth rate (CAGR) of 19.89% from 2024 to 2032 [73]. Continued advancements could enable BSi to improve energy recovery from industrial waste heat and support thermal energy conversion in space applications [74].

5.7 Summary

This chapter examines the thermal properties of BSi, including heat capacity, thermal conductivity, phonon scattering, and infrared absorption. It also explores temperature-dependent photoluminescence and applications in thermoelectric, photo-thermoelectric, and photovoltaic technologies, along with emerging uses in thermophotovoltaics, thermal sensing, infrared detectors, passive radiative cooling, and aerospace.

References

[1] Lv J *et al* 2018 Review application of nanostructured black silicon *Nanoscale Res. Lett.* **13** 110

[2] Liu X *et al* 2014 Black silicon: fabrication methods, properties and solar energy applications *Energy Environ. Sci.* **7** 3223–63

[3] Chai J Y H, Wong B T and Juodkazis S 2020 Black-silicon-assisted photovoltaic cells for better conversion efficiencies: a review on recent research and development efforts *Mater. Today Energy* **18** 100539

[4] Ayvazyan G *et al* 2022 Effect of thermal annealing on the structural and optical properties of black silicon *J. Mater. Sci. Mater. Electron.* **33** 17001–10

[5] Soueiti J *et al* 2023 A review of cost-effective black silicon fabrication techniques and applications *Nanoscale* **15** 4738–61

[6] Fan Z *et al* 2021 Recent progress of black silicon: from fabrications to applications *Nanomaterials* **11** 41

[7] Uddin S, Hashim M R and Pakhuruddin M Z 2021 Effects of annealing temperature towards properties of black silicon fabricated by aluminium-assisted chemical etching *Mater. Sci. Semicond. Process.* **133** 105932

[8] McConnell A and Goodson K E 2005 Thermal conduction in silicon micro-and nano-structures *Annu. Rev. Heat Transfer.* **14** 129–68

[9] Halliday D, Resnick R and Walker J 2013 *Fundamentals of Physics* (New York: Wiley)

[10] Ravindra N M, Marthi S R and Sekhri S 2016 *Black Silicon: Processing, Properties, and Applications* (New York: Momentum)

[11] Fourier J 1955 *Analytical Theory of Heat* (New York: Dover) p 359
[12] Zhang F and Gallagher K S 2016 Innovation and technology transfer through global value chains: evidence from China's PV industry *Energy Policy* **94** 191–203
[13] Yin X *et al* 2023 High thermoelectric performance of $TlInSe_3$ with ultra-low lattice thermal conductivity *Front. Phys.* **11** 1172989
[14] Li H *et al* 2022 Intrinsically and extrinsically anisotropic heat transport in bulk materials and nanostructures: a review *Int. J. Heat Mass Transfer* **196** 123307
[15] Zhang Z M, Zhang Z M and Luby 2007 *Nano/Microscale Heat Transfer* **vol 410** (Berlin: Springer)
[16] Fisher T S 2013 *Thermal Energy at the Nanoscale* **vol 3** (Singapore: World Scientific)
[17] Shin S *et al* 2020 Advanced materials for high-temperature thermal transport *Adv. Funct. Mater.* **30** 1904815
[18] Chen G 2005 *Nanoscale Energy Transport and Conversion: a Parallel Treatment of Electrons, Molecules, Phonons, and Photons* (Oxford: Oxford University Press)
[19] Klitsner T and Pohl R O 1987 Phonon scattering at silicon crystal surfaces *Phys. Rev.* B **36** 6551–65
[20] Tien C L 1997 *Microscale Energy Transfer* (Boca Raton, FL: CRC Press)
[21] Feng T, Lindsay L and Ruan X 2017 Four-phonon scattering significantly reduces intrinsic thermal conductivity of solids *Phys. Rev.* B **96** 161201
[22] Cahill D G *et al* 2003 Nanoscale thermal transport *J. Appl. Phys.* **93** 793–818
[23] Sojo Gordillo J M *et al* 2023 Recent advances in silicon-based nanostructures for thermoelectric applications *APL Mater.* **11** 040702
[24] Blatt F J 2020 *Matthiessen's Rule* (New York: McGraw-Hill)
[25] Wang Y, Lu Z and Ruan X 2016 First principles calculation of lattice thermal conductivity of metals considering phonon–phonon and phonon–electron scattering *J. Appl. Phys.* **119** 225109
[26] Tong Z *et al* 2019 Comprehensive first-principles analysis of phonon thermal conductivity and electron–phonon coupling in different metals *Phys. Rev.* B **100** 144306
[27] Liao B *et al* 2015 Significant reduction of lattice thermal conductivity by the electron–phonon interaction in silicon with high carrier concentrations: a first-principles study *Phys. Rev. Lett.* **114** 115901
[28] Devanathan V 2021 The Wedemann–Franz law for electrical and thermal conduction in metals *J. Chennai Acad. Sci.* **4** 1–26
[29] Lee H *et al* 2010 Effects of nanoscale porosity on thermoelectric properties of SiGe *J. Appl. Phys.* **107** 094308
[30] Yang H-Y *et al* 2020 Ultra-low thermal conductivity of roughened silicon nanowires: role of phonon–surface bond order imperfection scattering *Chin. Phys.* B **29** 086502
[31] Hao Q, Xiao Y and Zhao H 2016 Characteristic length of phonon transport within periodic nanoporous thin films and two-dimensional materials *J. Appl. Phys.* **120** 065101
[32] Minnich A *et al* 2009 Modeling study of thermoelectric SiGe nanocomposites *Phys. Rev.* B **80** 155327
[33] Ohishi Y *et al* 2015 Thermoelectric properties of heavily boron- and phosphorus-doped silicon *Jpn. J. Appl. Phys.* **54** 071301
[34] Yang L *et al* 2021 High thermoelectric figure of merit of porous Si nanowires from 300 to 700 K *Nat. Commun.* **12** 3926

[35] Ravichandran J *et al* 2014 Crossover from incoherent to coherent phonon scattering in epitaxial oxide superlattices *Nat. Mater.* **13** 168–72

[36] Balandin A A 2011 Thermal properties of graphene and nanostructured carbon materials *Nat. Mater.* **10** 569–81

[37] Möller H J *et al* 2005 Multicrystalline silicon for solar cells *Thin Solid Films* **487** 179–87

[38] Uma S *et al* 2001 Temperature-dependent thermal conductivity of undoped polycrystalline silicon layers *Int. J. Thermophys.* **22** 605–16

[39] Sverdrup P G, Sungtaek Ju Y and Goodson K E 2001 Sub-continuum simulations of heat conduction in silicon-on-insulator transistors *J. Heat Transfer* **123** 130–7

[40] Asheghi M *et al* 1998 Temperature-dependent thermal conductivity of single-crystal silicon layers in SOI substrates *J. Heat Transfer* **120** 30–6

[41] Ju Y S and Goodson K E 1999 Phonon scattering in silicon films with thickness of order 100 nm *Appl. Phys. Lett.* **74** 3005–7

[42] Asheghi M *et al* 2002 Thermal conduction in doped single-crystal silicon films *J. Appl. Phys.* **91** 5079–88

[43] Goodson K E *et al* 1995 Prediction and measurement of temperature fields in silicon-on-insulator electronic circuits *J. Heat Transfer* **117** 574–81

[44] Cheng S-Q, Cai B and Zhu Y-M 2015 Black silicon as absorber for near-infrared photothermal conversion *2015 Opto-Electronics and Communications Conf. (OECC)* (Piscataway, NJ: IEEE)

[45] Jiang P, Qian X and Yang R 2018 Tutorial: time-domain thermoreflectance (TDTR) for thermal property characterization of bulk and thin film materials *J. Appl. Phys.* **124** 161103

[46] Ould-Abbas A, Bouchaour M and Sari N-E C 2012 Study of thermal conductivity of porous silicon using the micro-Raman method *Open J. Phys. Chem.* **2** 1–6

[47] Holland M 1963 Analysis of lattice thermal conductivity *Phys. Rev.* **132** 2461

[48] Kittel C and McEuen P 2018 *Introduction to Solid State Physics* (New York: Wiley)

[49] Majumdar A 1993 Microscale heat conduction in dielectric thin films *J. Heat Transfer* **115** 7–16

[50] Li D *et al* 2003 Thermal conductivity of individual silicon nanowires *Appl. Phys. Lett.* **83** 2934–6

[51] Touloukian Y S *et al* 1971 Thermal Conductivity—Nonmetallic Solids Thermophysical Properties of Matter—The TPRC Data Series **vol 2** (New York: IFI/Plenum)

[52] Zhao Z *et al* 2023 Black silicon for near-infrared and ultraviolet photodetection: a review *APL Mater.* **11** 021107

[53] Sarkar S *et al* 2023 Black silicon revisited as an ultrabroadband perfect infrared absorber over 20 μm wavelength range *Adv. Photonics Res.* **4** 2200223

[54] Mi G *et al* 2023 Improvement in near-infrared absorbance attenuation by using nanometer black silicon composited with gold nanoparticles *Discover Nano* **18** 82

[55] Sarkar S *et al* 2019 NIR and MIR absorption of ultra-black silicon (UBS). Application to high emissivity, all-silicon, light source *IEEE 32nd Int. Conf. on Micro Electro Mechanical Systems (MEMS)* (Piscataway, NJ: IEEE)

[56] Moore A L and Shi L 2014 Emerging challenges and materials for thermal management of electronics *Mater. Today* **17** 163–74

[57] Serpengüzel A *et al* 2008 Luminescence of black silicon *J. Nanophotonics* **2** 021770

[58] Lü Q *et al* 2013 Strong infrared photoluminescence from black silicon made with femtosecond laser irradiation *Opt. Lett.* **38** 1274–6

[59] Stranz A *et al* 2011 Measurements of thermoelectric properties of silicon pillars *Sensors Actuators* A **171** 48–53

[60] Ju Y S 2005 Phonon heat transport in silicon nanostructures *Appl. Phys. Lett.* **87** 153106

[61] Martin P *et al* 2009 Impact of phonon-surface roughness scattering on thermal conductivity of thin Si nanowires *Phys. Rev. Lett.* **102** 125503

[62] Huang I-Y *et al* 2008 Development of low-cost micro-thermoelectric coolers utilizing MEMS technology *Sensors Actuators* A **148** 176–85

[63] Hwang G C *et al* 2018 Pressure-driven phase transitions and reduction of dimensionality in 2D silicon nanosheets *Nat. Commun.* **9** 5412

[64] Zhou Y and Hu M 2016 Record low thermal conductivity of polycrystalline Si nanowire: breaking the Casimir limit by severe suppression of propagons *Nano Lett.* **16** 6178–87

[65] Donadio D and Galli G 2009 Atomistic simulations of heat transport in silicon nanowires *Phys. Rev. Lett.* **102** 195901

[66] Zhao Z *et al* 2024 Enhanced photo-thermoelectric detection in black silicon with chimney-like texture *Mater. Today Commun.* **40** 109723

[67] Hervé A *et al* 2023 Passive nighttime radiative cooling using black silicon *Proc. of the 10th Int. Symp. on Radiative Transfer, RAD-23 (Thessaloniki, Greece, 12–16 June 2023)* (Danbury, CT: Begel House)

[68] Zhong S *et al* 2012 The study on the properties of black multicrystalline silicon solar cell varying with the diffusion temperature *Energy Procedia* **14** 505–11

[69] Khairul Azri A A *et al* 2025 Development of surface-texturized black silicon through metal-assisted chemical etching and its application in the thermophotovoltaic field: a review and recommendation *Semicond. Sci. Technol.* **40** 013001

[70] Bardalen E *et al* 2024 Black silicon as anti-reflective structure for infrared imaging applications *Nanomaterials* **14** 20

[71] Georgiev G T *et al* 2019 The bidirectional reflectance of black silicon used in space and earth remote sensing applications *Proc. SPIE* **11151** 111511E

[72] Ron S *et al* 2021 Black-silicon as a wideband infrared absorber for the space instruments: fabrication, modeling, and characterization *Proc. SPIE* **11819** 118190N

[73] Market Research Intellect 2025 Black silicon technology market size, share & trends by product, application & geography—Forecast to 2033 *Report* 998838 Market Research Intellect https://www.marketresearchintellect.com/product/global-black-silicon-technology-market/

[74] Datas A and Marti A 2017 Thermophotovoltaic energy in space applications: review and future potential *Sol. Energy Mater. Sol. Cells* **161** 285–96

Chapter 6

Electronic properties

This chapter explores the fundamental electronic properties of nanostructured black silicon (BSi), including a brief overview of the carrier transport mechanisms and electrical conductivity. Important parameters such as effective carrier lifetime and surface recombination velocity are examined to understand the electronic properties. The chapter also presents analyses of the band structure and density of states, highlighting their impact on the device performance and expanding the application range of BSi.

6.1 Introduction

Black silicon (BSi) is a nanostructured variant of silicon featuring high surface roughness and enhanced optical absorption due to its unique morphology. Its nanoscale texture significantly lowers the reflectivity by 96%–98% across the visible and infrared (IR) wavelengths, making it highly efficient for solar cells and photodetectors [1]. The exceptional light-trapping capability of BSi boosts energy conversion efficiency in optoelectronic devices. However, the increased surface area results in higher recombination rates, necessitating effective passivation to sustain device performance. Additionally, BSi can accommodate high concentrations of dopants such as sulfur, which modify its electronic properties [2]. These properties make BSi a promising material for improving solar cell efficiency and enabling optical detection with high sensitivity.

While primarily studied for its optical characteristics, the electronic properties are equally important for its performance in semiconductor applications. Although BSi retains the fundamental electronic properties of crystalline silicon (c-Si), nanostructuring introduces surface defects that create mid-gap states, leading to reduced mobility and increased recombination [3–5]. Charge carrier behavior is influenced by band bending and trap-assisted transport, while doping efficiency and contact resistance are affected by surface roughness. Its applications span photovoltaics,

doi:10.1088/978-0-7503-3247-7ch6

photodetectors, and sensors [1], where optimizing the passivation and doping strategies are essential for achieving high device efficiency.

6.2 Fundamentals

The electronic properties of a material determine its response to electric fields, charge carriers, and external stimuli, influencing its applications in semiconductors, photovoltaics, and electronic devices [3–6]. These properties play a key role in distinguishing conductors, insulators, and semiconductors, as they govern heat flow, electric currents, and overall material behavior, making electronic interactions fundamental to most material characteristics.

Electronic properties are closely linked to the microstructure of the material, which consists of phases, defect structures, and residual stress, along with their distributions. Thermo-mechanical treatments can be used to tailor the microstructure, enabling the design of materials with specific properties. In practical applications, hybrid systems often integrate multiple materials with distinct microstructures to optimize performance for specialized functions [3–6].

Microstructural features and electronic transport phenomena also influence stress and strain within materials. Residual stress, which exists without applied external forces, can appear at macro- or micro-scales and may have either beneficial or detrimental effects. Additionally, interfaces act as mirrors for electrons, necessitating atomic-level precision due to the short electron wavelengths [3–6].

Important electronic properties such as carrier lifetimes (τ), surface recombination velocities (S), density of states (DOS), and band structure are essential for understanding and predicting material behavior in various technological applications.

6.2.1 Comparison to bulk silicon

BSi differs from bulk c-Si in its optical, electronic, and surface properties, offering distinct advantages for applications in photovoltaics, photodetectors, and sensing. Its nanostructured surface dramatically reduces reflectivity, leading to enhanced light absorption. Although both BSi and c-Si share a similar bandgap of approximately 1.12 eV, the increased density of defect states in BSi enhances Shockley–Read–Hall (SRH) recombination, resulting in reduced carrier lifetimes. Moreover, surface roughness and defects contribute to lower carrier mobility, impacting charge transport. Charge trapping further leads to band bending, altering carrier distribution and electrical conductivity. These unique properties make BSi well-suited for specialized optoelectronic and sensing applications [3, 7]. Table 6.1 presents a comparison of the fundamental material properties of bulk c-Si and BSi, focusing on the bandgap, density of states, carrier mobility, and surface recombination characteristics.

6.2.2 Carrier transport mechanisms

Carrier transport in BSi is inherently complex due to its nanostructured morphology, surface defects, and doping-related effects [10]. Several transport mechanisms

Table 6.1. Important electronic properties of bulk c-Si and BSi, including the bandgap, density of states, carrier mobility, and surface recombination [1, 3, 5, 7–9].

Property	Crystalline silicon (c-Si)	Black silicon (BSi)
Bandgap [1, 3, 7]	~1.12 eV (indirect)	Slightly modified due to surface effects
Density of states [3, 7]	Well-defined near-band edges	Additional mid-gap states from defects
Carrier mobility [9]	High	Reduced due to increased scattering
Surface recombination [1, 5, 8]	Moderate with passivation	High, requiring effective passivation

contribute to charge movement, including band transport, hopping conduction, trap-assisted transport, and tunneling. In high-quality BSi, carriers primarily move through the conduction and valence bands; however, nanostructuring, surface states, and mid-gap defects introduce significant scattering, reducing carrier mobility. Additionally, heavy doping alters the band structure, further impacting both mobility and carrier concentration.

Deep-level defects and surface states act as traps, causing non-radiative recombination and lowering charge collection efficiency. Mid-gap states increase SRH recombination, resulting in shorter carrier lifetimes. Surface passivation techniques (disused in chapter 8) such as using aluminum oxide (Al_2O_3) or silicon nitride (SiN_x) are essential to mitigate these losses by reducing dangling bonds [3, 5, 8]. While Auger recombination is generally unaffected, it can increase with high doping [11]. Effective passivation is therefore crucial to minimize the overall recombination losses [12, 13].

In heavily doped BSi, charge transport often occurs via hopping conduction between localized states. Phonon-assisted hopping dominates at moderate temperatures, while variable range hopping (VRH), enabled by quantum tunneling, becomes prominent at lower temperatures. Sulfur doping reduces dopant activation energy, thereby stabilizing carrier concentration across a wide range of temperatures [14].

Overall, carrier transport in BSi depends on surface states, doping, and the environment. Band transport dominates in well-passivated structures, while traps, hopping, and tunneling prevail in defective or heavily doped regions. Optimized passivation and interface engineering is crucial to advancing BSi-based devices.

6.2.3 Doping and electrical conductivity

Doping in BSi follows the foundational principles of conventional silicon, but its nanoscale morphology introduces distinct challenges and opportunities. Doping plays a critical role in controlling the electrical conductivity of BSi, directly impacting carrier concentration, transport behavior, and overall device performance in electronic and optoelectronic applications. The nanostructured surface of BSi modifies dopant activation and recombination processes compared to bulk silicon [15].

Nanostructuring affects the activation energy of dopants, often resulting in incomplete ionization at room temperature. BSi can accommodate high dopant levels of, for example, sulfur to boost conductivity, but excessive doping may increase recombination, reduce mobility, and cause bandgap narrowing. While sulfur doping can enable temperature-insensitive hopping conduction, it may also negatively impact carrier dynamics.

Carrier mobility in BSi is generally lower than in bulk silicon due to various scattering mechanisms [9] as discussed in chapter 5. Surface roughness intensifies electron–phonon interactions, while defects create localized trap states that promote SRH recombination. Additionally, high doping concentrations introduce Coulombic scattering, further hindering carrier movement.

The rough surface morphology of BSi also contributes to increased contact resistance, complicating the formation of low-resistance metal-semiconductor junctions. In order to address this, strategies such as engineered metal contacts, heavily doped surface regions, and conductive coatings are employed. In highly doped BSi, quantum tunneling effects, such as Fowler–Nordheim tunneling [16], can facilitate enhanced carrier injection and are particularly beneficial for photodetectors and sensor applications [15]. Owing to the increased surface area, achieving low-resistance ohmic contacts requires optimized metal–silicon interfaces and passivation layers.

6.3 Lifetime and surface recombination velocity

BSi generally exhibits shorter carrier lifetimes compared to bulk c-Si, mainly due to enhanced recombination from surface defects and mid-gap states. However, effective surface passivation can significantly mitigate these recombination losses [3–5, 9].

The strong near-band-edge absorption in BSi arises from its nanostructured surface, which enhances light trapping through multiple internal reflections and scattering. Deep BSi structures offer superior optical performance due to steeper sidewalls and optimized spacing (chapter 7), while shallow and intermediate structures exhibit similar behavior owing to comparable sidewall angles. When fabrication minimizes plasma-induced damage and effective passivation is applied, BSi can achieve carrier lifetimes that are comparable to those of polished silicon wafers [17]. However, surface structuring increases the effective surface area and introduces fabrication-induced defects, both of which contribute to elevated surface recombination rates and reduced near-surface bulk lifetime [18, 19]. As a result, BSi-based solar cells often exhibit lower efficiencies than state-of-the-art commercial solar cells [18, 20–22]. As discussed in chapter 8, the effective minority carrier lifetime (τ_{eff}) in such structures can be modeled as the combined effect of bulk and surface contributions, described by the following relation [23–25]:

$$\frac{1}{\tau_{\text{eff}}} = \frac{1}{\tau_{\text{bulk}}} + \frac{S_{\text{front}} + S_{\text{back}}}{W}, \tag{6.1a}$$

where τ_{bulk} is the bulk lifetime, and S_{front} and S_{back} are the surface recombination velocities at the front and back surfaces, respectively. Assuming $\tau_{eff} \ll \tau_{bulk}$, for a non-textured wafer $S_{front} = S_{back} = S_{planar}$, the effective lifetime can be written as

$$\frac{1}{\tau_{eff}} = \frac{2S_{planar}}{W}. \tag{6.1b}$$

Therefore, S_{planar} can be written as

$$S_{planar} = \frac{W}{2\tau_{eff}}. \tag{6.2a}$$

The surface recombination velocity ($S_{textured}$) for one-sided textured wafers is calculated using the formula

$$S_{textured} = \frac{W}{\tau_{eff}} - S_{planar}. \tag{6.2b}$$

Figure 6.1 presents the carrier lifetimes of shallow, intermediate, and deep BSi structures, along with a polished reference (all coated with 100 nm of Al_2O_3), measured using the quasi-steady-state-photoconductance (QSSPC) technique after annealing [18]. At an excess carrier density of $\Delta N = 5 \times 10^{15}$ cm^{-3}, the intermediate BSi structure and the reference sample showed high effective lifetimes (1.4 and 1.6 ms, respectively). In contrast, the shallow and deep BSi structures exhibited lower lifetimes (500 and 230 μs) (table 6.2). Surface recombination velocities are calculated assuming an infinite bulk lifetime (equation (6.2a)) [25].

Table 6.2 presents summaries of τ_{eff} and S_{eff} for various Al_2O_3-passivated BSi structures. Table 6.2 highlights the influence of surface structuring on the electronic properties of BSi. Variations in lifetime and recombination velocity reflect how nanostructuring impacts the charge carrier dynamics which directly impacts solar cell efficiency [18].

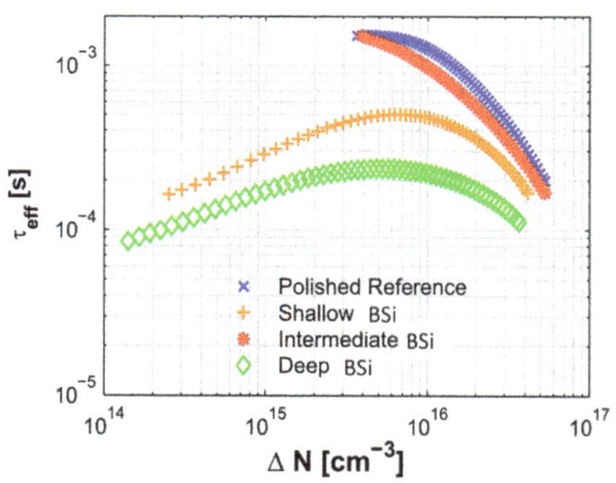

Figure 6.1. Injection-level dependent lifetimes of shallow, intermediate, and deep BSi structures, along with a polished reference sample passivated with 100 nm of Al_2O_3. All samples are annealed in a low-pressure argon ambient at 425 °C for 30 min. (Reproduced with permission from [18]. Copyright 2012 AIP Publishing.)

Table 6.2. Summary of effective carrier lifetimes (τ_{eff}) and corresponding surface recombination velocities (S_{eff}) of Al_2O_3-passivated BSi structures as shown in figure 6.1. These are measured at an excess carrier density of $\Delta N = 5 \times 10^{15}$ cm^{-3}, assuming an infinite bulk lifetime [18].

Structure	Material; Czochralski (CZ)	τ_{eff} (μs)	S_{eff} (cm s^{-1})
Planar structure	CZ	1634 ± 245	12 ± 2
Shallow BSi	CZ	502 ± 76	45 ± 7
Intermediate BSi	CZ	1475 ± 133	13 ± 2
Deep BSi	CZ	237 ± 36	80 ± 6

A study has shown the value of the surface recombination velocity of ~6.9 cm s^{-1} and an open-circuit voltage (V_{oc}) of ~757 mV for Al_2O_3-coated p-type CZ wafers (180 μm thick, 4 Ω·cm base resistivity). The experiment involved a non-cryogenic inductively coupled plasma reactive ion etching (ICP-RIE) technique for silicon surface texturing that minimizes damage and enhances BSi solar cell performance [26]. This method achieved ultra-low reflectance—below 3% across the 300–1000 nm wavelength range within just 2 min of etching [26].

Figure 6.2(a) presents τ_{eff} as a function of reactive ion etching time (t_{RIE}), measured over a 100 mm wafer area. Lifetime measurements were conducted within an hour after post-atomic layer deposition (ALD) annealing and monitored over approximately 300 h to evaluate passivation stability. Although τ_{eff} initially degraded due to likely water intercalation during wafer storage, it stabilized within 200 h and could be fully recovered through 10 min of hotplate annealing at 150 °C. Similar degradation effects have been reported previously [27–29], which can be mitigated by applying a SiN_x:H capping layer.

Immediately after post-ALD annealing, τ_{eff} exceeded 3 ms for t_{RIE} between 1.5 and 2 min, and between 2.5 and 3 ms for t_{RIE} of 3 min. In comparison, non-textured wafers showed a τ_{eff} of 3.7 ms, while texturing with capacitively coupled plasma (CCP) significantly reduced τ_{eff} to 0.89 ms. For t_{RIE} = 3 min, τ_{eff} remained stable above 1.5 ms, although further increases in etching time resulted in declining lifetime values. These results suggest that 2–3 min of ICP-RIE offers an optimal trade-off between reflectance reduction and minimal surface damage [26].

Figure 6.2(b) shows the calculated values of $S_{textured}$ using equation (6.2b). For t_{RIE} of 1.5, 2, and 3 min, $S_{textured}$ remained below 10 cm s^{-1} (6.9, 8.7, and 9.7 cm s^{-1}, respectively), only slightly higher than the non-textured wafer (6.8 cm s^{-1}). While longer t_{RIE} marginally increased $S_{textured}$, the change was not substantial. In contrast, CCP texturing resulted in a five-fold increase, reaching 49.4 cm s^{-1}. Stabilized τ_{eff} data yielded extracted $S_{textured}$ values of 12.9 and 13.6 cm s^{-1} for t_{RIE} of 1.5 and 2 min-values close to the S_{planar} (10.6 cm s^{-1}).

Notably, ICP-RIE texturing for 3 min or shorter results in $S_{textured}$ values (before degradation) that are on par with or even exceed the state-of-the-art for p-type CZ wafers, with values of 11 cm s^{-1} as reported by Otto *et al* [30], 20 cm s^{-1} in the record BSi cells by Savin *et al* [31], and 10 cm s^{-1} as reported by Allen *et al* [32]. This highlights its potential for high-efficiency silicon solar cell fabrication.

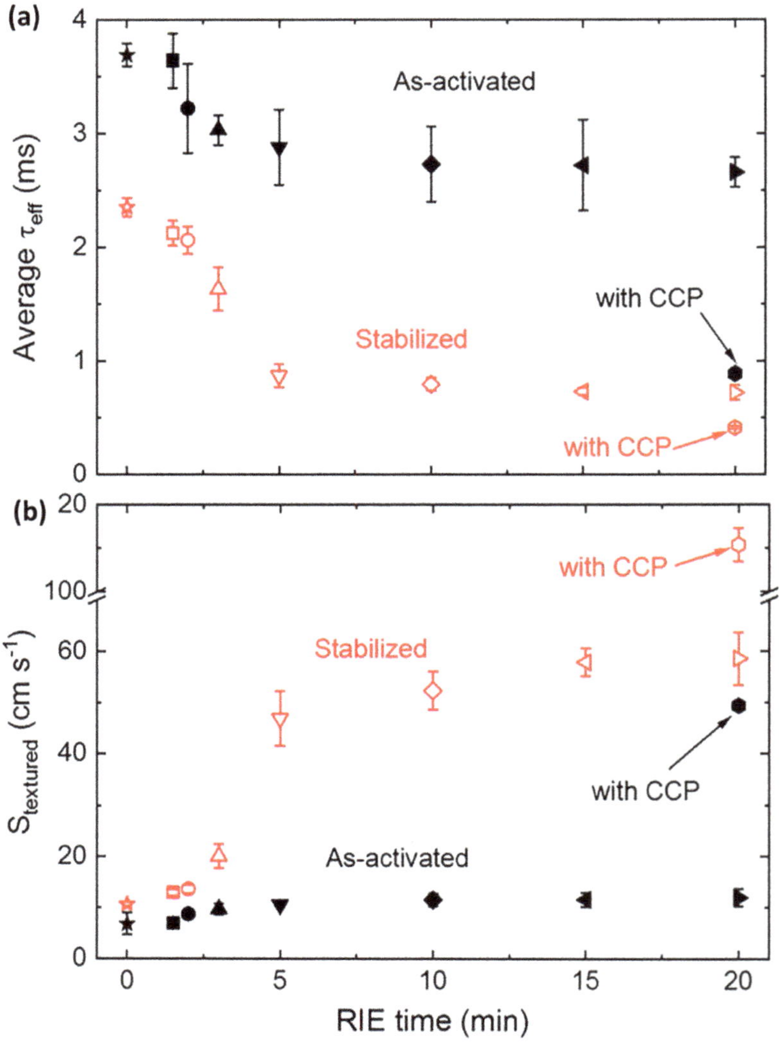

Figure 6.2. (a) The average values of minority carrier effective lifetime (τ_{eff}), as a function of etching time (t_{RIE}), are mapped across 100 mm diameter surfaces, including a wafer textured using CCP. Error bars represent one standard deviation from the average. (b) The average effective surface recombination velocity of the textured surface ($S_{textured}$) is shown, with error bars indicating one standard deviation from the average. Open symbols correspond to values measured within one hour of Al_2O_3 activation following post-ALD annealing, while closed symbols represent values measured after surface passivation stabilization. (Reproduced with permission from [26]. Copyright 2019 Wiley.)

6.4 Band structure and density of states

The electronic band structure of BSi closely resembles that of c-Si. Nonetheless, its unique surface morphology and nanostructuring introduce significant modifications. Localized defect states often emerge, creating mid-gap states in the DOS, which in turn degrades performance in optoelectronic applications. In structures with high aspect ratios, quantum confinement effects may cause a slight widening of the bandgap. Meanwhile, surface states can alter the band edges, further influencing the density of free carriers [7, 23, 33].

6.4.1 Band structure

The band structure of a material defines the allowed energy levels for electrons, comprising the valence band (V_B), where electrons are bound to atoms, and the conduction band (C_B), where electrons can move freely. The energy gap between these bands, known as the bandgap (E_g), governs the electrical behavior of materials. Semiconductors possess relatively small band gaps, allowing them to absorb the incident light and promote electrons from the V_B to the C_B, while insulators have wide gaps that inhibit electrical conduction. The Fermi level (E_F), which marks the energy level with a 50% probability of electron occupancy, lies between the V_B and C_B in intrinsic semiconductors [7, 33].

In doped semiconductors, the type and concentration of dopants determine the distribution of charge carriers, maintaining electrical neutrality through a balance of localized dopant ions and mobile opposite charges. In p-type silicon, doped with Group III elements such as boron, the majority carriers are holes, created when acceptor atoms capture electrons from the valence band [34]. Each hole is paired with a negatively charged dopant ion to preserve charge neutrality. In this case, E_F shifts closer to the valence band (figure 6.3(a)) and the surface charge distribution forms a repeating pattern of negative dopant ions surrounded by positive delocalized holes (figure 6.3(b)) [34].

Conversely, in *n*-type silicon, Group V elements such as phosphorus act as donors, contributing free electrons to the conduction band. When a donor atom donates an electron, it creates a delocalized negative charge cloud around the positively charged dopant ions. The majority carriers, free electrons, accumulate near the surface, and E_F shifts closer to the conduction band (figure 6.3(a)). The corresponding surface charge distribution consists of negatively charged electron regions surrounding the positively charged donor ions, as shown in figure 6.3(b) [34].

The mean dopant spacing, denoted by λ, represents the average distance between adjacent point charges, and is directly related to the dopant density (N_D) of the doped silicon material [35, 36]

$$\lambda = N_D^{\ (-1/3)}. \tag{6.3}$$

Thus, dopant density plays a crucial role in shaping the overall surface potential corrugation and electrical properties of doped silicon.

A study reports variations in band structure and DOS with varying sulfur doping. Using femtosecond laser pulse techniques, the researchers fabricated BSi micro-structured arrays consisting of sharp conical spikes approximately 40 μm in height [37]. The electronic band structures corresponding to varying sulfur concentrations are shown in figure 6.4 [37] depicting the energy dispersion characteristics of sulfur-doped silicon.

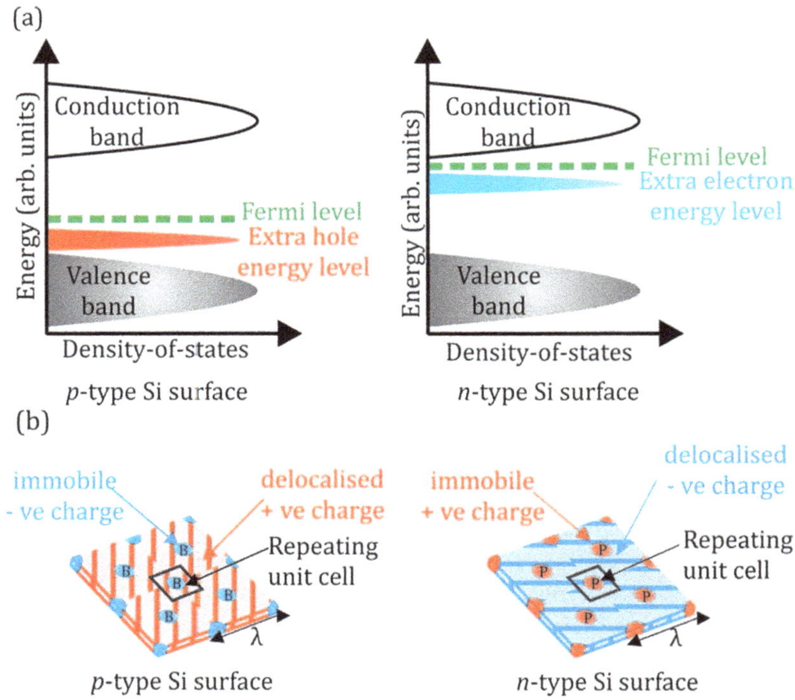

Figure 6.3. (a) Schematic diagram showing the DOS for p-type and n-type doped silicon surfaces. The shaded region represents the energy levels occupied below the Fermi level at 0 K. (b) A model of p-type and n-type doped silicon surfaces, showing the distribution of surface charges required to maintain electrical neutrality due to the presence of point-charged dopants. The parameter λ represents the mean dopant spacing, although in reality, dopant spacing is irregular, despite being modeled as uniform. (Reproduced with permission from [34]. Copyright 2013 AIP Publishing.)

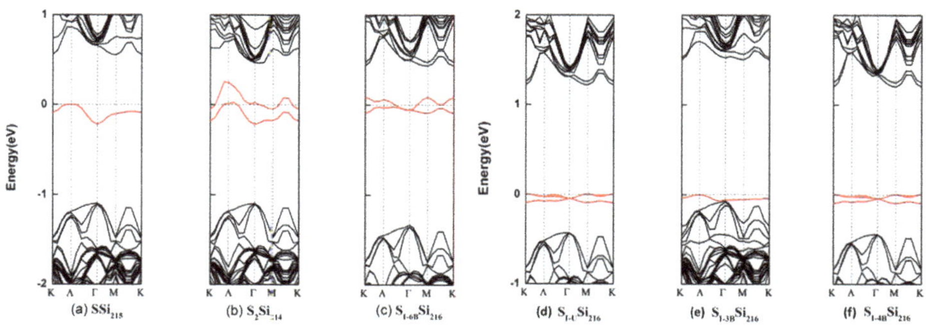

Figure 6.4. Electronic band structures of BSi with different sulfur doping configurations. The red lines represent the impurity bands introduced by sulfur incorporation and the dotted line at 0 eV indicates the Fermi level. The k-point coordinates in the Brillouin zone (BZ) are defined as K [0,0,0.5], Λ [0,0.5,0], Γ [0,0,0], and M [0,0.5,0.5]. (Reproduced with permission from [37]. Copyright 2015 American Chemical Society.)

The band structure plots reveal the emergence of impurity-induced states near E_F, highlighted in red, which indicate enhanced electron delocalization. As seen in figures 6.4(a), 6.4(b), and 6.4(c), a narrow impurity band (~0.22 eV) emerges approximately 0.51 eV below the conduction band edge. These impurity bands span across E_F and shift closer to the conduction band with increasing sulfur concentration. The broadening and fluctuation of these bands suggest an increase in electron delocalization, characteristic of a transition toward metallic behavior.

As doping levels rise, the Fermi level progressively approaches the conduction band edge. At high sulfur concentrations, not all dopant atoms are substitutionally integrated into the silicon lattice, as inferred from the band structure and formation energy analyses. With increased sulfur content, the system exhibits intensified breakdown effects, culminating in an insulator-to-metal transition within the concentration range of 0.4%–0.9% [14]. At these elevated doping levels, the emergence of denser energy bands signals enhanced metallic characteristics and improved conductivity, driven by bandgap narrowing and stronger electron–electron interactions.

Figures 6.4(d), 6.4(e), and 6.4(f) illustrate the band structures of models where sulfur atoms occupy interstitial sites in the silicon lattice, specifically $S_{I-U}Si_{108}$, $S_{1\text{-}2}BSi_{216}$, and $S_{I-I}Si_{108.}$ In these configurations, impurity bands are absent. However, a distinct energy band appears near the valence band edge, indicating electrons in metastable states. Variations in doping conditions lead to the formation of deep donor or bound states, that are attributed to unpaired electrons. A narrow impurity band (~0.09 eV) appears approximately 0.39 eV above the valence band edge, corresponding to shallow donor levels.

In BSi systems, the electronic properties are predominantly influenced by the spatial positioning of dopant atoms rather than their bonding configurations [37]. Impurity levels located near the conduction band edge are primarily responsible for the observed metallicity. Thus, both dopant concentration and atomic positioning play crucial roles in defining the behavior of sulfur-doped BSi.

6.4.2 Density of states

The DOS describes the number of available electronic states at each energy level within a material, playing a pivotal role in determining its electronic and optical properties. Commonly illustrated as a plot of DOS versus energy, it reflects the distribution of states that electrons can occupy. In semiconductors and insulators, the DOS near the valence and conduction band edges is critical, as it governs the electron excitation and transport mechanisms. The valence band DOS comprises states that are naturally occupied by electrons, while the conduction band DOS consists of states that are accessible to electrons upon gaining sufficient energy. E_F represents the highest occupied energy level at absolute zero. The value of the DOS at E_F directly influences the electrical conductivity. Metals have a high DOS at E_F, allowing for free electron movement and resulting in high conductivity. Semiconductors display a lower DOS near E_F, which increases with distance from the band edges, influencing charge transport efficiency. In insulators, the DOS at E_F is nearly zero due to the large bandgap, preventing electron flow [7, 23, 33].

The DOS also significantly affects optical absorption and electronic transport, determining the availability of states for electron excitation. In semiconductors such as perovskites, the DOS governs charge carrier generation, recombination, and mobility. Optimizing the DOS near the band gap is key to enhancing charge carrier separation and improving the efficiency of solar cells [7, 23, 33].

In BSi, the DOS deviates from that of bulk c-Si due to factors such as surface defects, nanostructuring, and surface band bending. Surface band bending can also lead to the formation of depletion layers, altering charge distribution and limiting conductivity—particularly in undoped or lightly doped regions [37].

The DOS can be derived from the band structure of a material, and its form depends on the dimensionality (1D, 2D, or 3D) and the shape of the energy bands. For example, DOS in a 3D semiconductor is expressed as

$$g(E) = \frac{(2m^*)^{3/2}}{2\pi^2\hbar^3}\sqrt{E - E_c}, \tag{6.4}$$

where $g(E)$ is the density of states (per unit volume per energy), m^* is the effective mass of the charge carriers, $\hbar$ is the reduced Planck's constant ($\hbar = h/2\pi$), E is the energy, and E_c is the conduction band edge energy. This formula is derived from the free-electron approximation in a parabolic band structure [7, 23, 33].

Figures 6.5(a)–e) present the partial DOS (PDOS) for SSi_{215}, S_2Si_{214}, $S_{I\text{-}U}Si_{216}$, $S_{I\text{-}4B}Si_{216}$, and $S_{I\text{-}6B}Si_{216}$, highlighting the impact of sulfur on the electronic structure of silicon. Configurations such as, $S_{I\text{-}U}Si_{108}$, $S_{I\text{-}2B}Si_{216}$, $S_{I\text{-}3B}Si_{216}$, and $S_{I\text{-}I}Si_{108}$ lack an intermediate band. In figure 6.5(a), prominent Si p and S s electron contributions near the E_F suggest the formation of Si–S hybridized σ bands. Figure 6.5(b) shows a broad DOS peak at E_F, indicating that electrons in these bands are no longer bound. The absence of S–S covalent bonds implies the presence of coordination bonds instead [37].

The PDOS analysis for $S_{I-U}Si_{216}$ and $S_{I\text{-}4B}Si_{216}$ (figures 6.5(c) and (d)) reveals distinct electronic characteristics. In figure 6.5(c), a prominent peak at approximately 0 eV in the pink line indicates that the impurity bands near E_F are entirely composed of S p electrons. Similarly, figure 6.5(d) shows a DOS peak near E_F, also dominated by S p electrons; however, an additional peak at −0.4 eV suggests a relatively low Si–S bond energy. Figure 6.5(e) represents the PDOS for $S_{I\text{-}6B}Si_{216}$, where DOS peaks appear on both sides of E_F, indicating strong Si–S covalent bonding. However, antibonding interactions between Si p and S p orbitals are also present, further influencing the electronic structure [37].

Structural and PDOS analysis reveals that in $S_{I-U}Si_{216}$, the absence of direct Si–S bonds results in free electrons from sulfur enhancing conductivity. In contrast, $S_{I\text{-}4B}Si_{216}$ likely features ionic Si–S bonds, enabling electric dipole formation during vibrations and contributing to stronger below-band-gap absorption [37]. As sulfur concentration increases, impurity bands merge with the conduction band, closing the energy gap at the Fermi level and promoting electron delocalization. This leads to

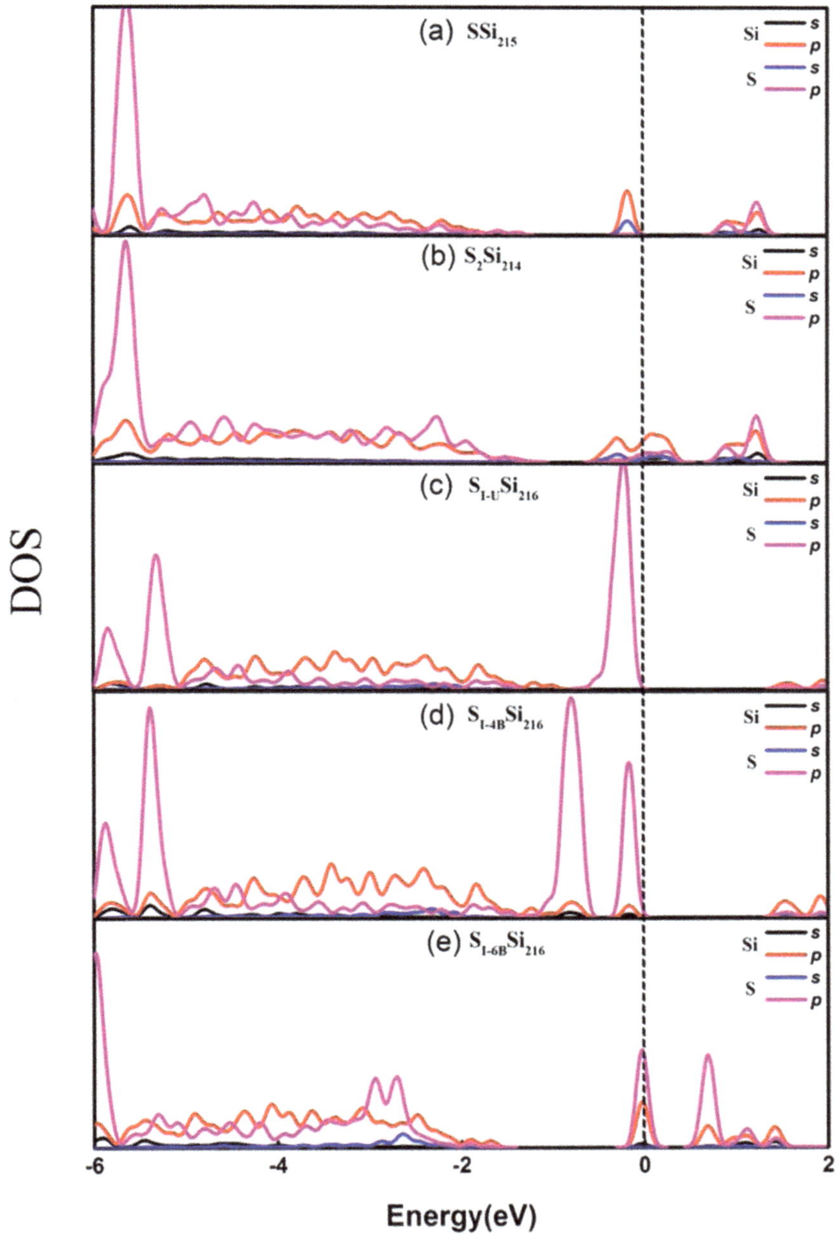

Figure 6.5. PDOS for different sulfur-doped silicon structures. Panels (a)–(e) represent various sulfur incorporation configurations, including substitutional and interstitial doping: (a) SSi_{215}, (b) S_2Si_{214}, (c) $S_{I-U}Si_{216}$, (d) $S_{I-4B}Si_{216}$, and (e) $S_{I-6B}Si_{216}$. The Si s, Si p, S s, and S p DOS are represented by black, red, blue, and pink lines, respectively. The vertical dashed line at 0 eV represents the Fermi level (E_F). The presence of mid-gap states and modifications to the conduction and valence band edges indicate the impact of sulfur doping on the electronic structure of silicon. (Reproduced with permission from [37]. Copyright 2015 American Chemical Society.)

stronger, more nonpolar Si–S interactions and intensified electron–electron interactions, further blurring the band gap. While free electrons boost infrared absorption, ultraviolet absorption remains limited due to energy overlap and reduced photon effectiveness. Additionally, increased recombination centers diminish overall absorption.

6.5 Applications and impact on device performance

BSi possesses unique electronic and optical properties that make it a highly versatile material across a wide range of applications [1, 3, 15]. Its enhanced light absorption significantly benefits photovoltaic devices, although effective surface passivation is essential to mitigate recombination losses caused by defects. In photodetectors and sensors, the nanostructured surface of BSi improves sensitivity—especially in the IR region. While c-Si remains the material of choice for high-performance electronics, BSi offers functional advantages due to its altered electronic structure. These include extended IR detection in photodetectors, enhanced surface interaction in chemical and biosensors, and modulation of voltage characteristics in solar cells. Expanding on the applications of BSi, covered in chapters 5, 7, 8, and 9, this section highlights its multifunctional potential, with a focus on the current and emerging uses related to its electronic properties.

In photovoltaics, BSi improves the solar cell efficiency significantly by increasing the light absorption and reducing the surface reflectivity to below 4%, thereby boosting the short-circuit current density (J_{sc}). However, the increased surface area introduces a high density of mid-gap states and lowers both τ and V_{oc}. Advanced passivation methods-such as ALD of Al_2O_3 and hydrogenation are essential to counter these effects [18]. When properly passivated, BSi solar cells can achieve efficiencies that are comparable to or exceeding those of traditional silicon cells, particularly under low-light or angled-light conditions [11, 38]. Due to these properties, BSi is ideal for high-efficiency solar applications such as building-integrated photovoltaics (BIPV), portable solar chargers, and cost-effective silicon-based panels.

In the realm of photodetectors and optical sensors, BSi extends spectral sensitivity into the infrared through its mid-gap states and efficient light-trapping effects, significantly enhancing the external quantum efficiency (EQE) in both UV and IR regions. Its high surface area and ability to undergo chemical functionalization make it especially well-suited for biosensors [39], gas sensors [40], and environmental monitors [41].

While surface defects can increase the dark current, optimized passivation, and doping strategies help maintain a high signal-to-noise ratio. BSi supports a variety of optoelectronic devices, including photodiodes, phototransistors, and quantum dot-based sensors. For example, gold-coated BSi nano-needles have been shown to penetrate bacterial membranes, enabling enhanced detection via surface-enhanced Raman spectroscopy (SERS)—a promising approach for lab-on-a-chip diagnostics [41].

Furthermore, BSi has been explored as an anti-reflective coating for silicon caps in vacuum-packaged microbolometer arrays, improving the sensitivity of IR detectors [39, 42–44]. Although its use in systems such as MRI is still under investigation, BSi shows promise for expanding medical imaging technologies.

In micro-electromechanical systems (MEMS), BSi improves the performance of microsensors and actuators [45], particularly enhancing the accuracy of devices such as pressure sensors. Its strong light absorption also benefits MEMS-based energy harvesters by facilitating efficient conversion of ambient light. Research shows that BSi can triple the signal-to-noise ratio in sensors and extend optical readout distances up to 12 cm. Additionally, its anti-biofouling properties suppress tissue adhesion and inflammatory responses, making it suitable for biomedical implantable sensors [46]. With appropriate engineering, BSi also holds the potential for power semiconductor devices despite its inherent surface defect challenges [30].

Its nanoscale surface topography improves performance in capacitive and piezoresistive sensors [47], while in wearable electronics, BSi facilitates ambient light harvesting for photovoltaic-powered devices such as smartwatches and fitness trackers. Its compatibility with flexible and lightweight substrates supports the development of self-sustaining wearable systems that can power low-energy sensors, displays, and communication components without the need for bulky batteries [48, 49].

From an electronics standpoint, the surface roughness of BSi and defect-induced scattering reduce carrier mobility, which can negatively impact transistor performance and increase contact resistance. Therefore, interface engineering, along with controlled doping and effective passivation, is essential for achieving reliable charge transport. While high doping levels can enable tunable electrical properties, excessive doping may degrade mobility and increase recombination. In imaging technologies, BSi improves absorption across the infrared, visible, and ultraviolet spectra, enhancing the performance of IR detectors used in thermal imaging and night vision—critical for both security and medical diagnostics. It also increases the image sensitivity and quality in scientific and consumer-grade cameras.

In summary, exceptional light absorption of BSi, tunable electrical properties, and nanoscale adaptability make it a promising material for a wide range of emerging technologies. Its ability to enhance solar energy conversion, optical sensing, MEMS, and imaging systems—as well as its growing role in biomedical and IoT applications—positions BSi as a breakthrough material for the next-generation electronic and optoelectronic devices.

A recent study [50] explores the light-emitting characteristics of BSi, enhancing its potential for use in efficient luminescent structures and the next-generation sensing technologies. The study investigates BSi fabricated via femtosecond laser ablation in air, highlighting its surface morphology, optical absorption, and carrier lifetime. It emphasizes the impact of an oxygen-rich surface layer on absorption and carrier dynamics, analysed through elemental composition and photoluminescence (PL) measurements [50].

An important parameter in this investigation is the empirical figure of merit ξ [51], which is proportional to the laser repetition rate and inversely proportional to the scan speed. By examining ξ, the study aims to understand its role in shaping the surface morphology and light absorptance, providing insights for optimizing the laser processing conditions for various BSi-based applications.

Originally introduced by Nava *et al* [51], the parameter ξ quantifies cumulative fluence and demonstrates a strong correlation with optical absorption. The present study extends this concept to explore how ξ influences additional surface characteristics, particularly luminescent features that are often studied independently [52, 53].

By systematically varying laser parameters such as repetition rate and scan speed, a wide range of ξ values are generated, enabling the fine-tuning of laser settings for specific application requirements. The parameter ξ (in J cm^{-2}) is defined as the cumulative laser energy density over a given area:

$$\xi = \frac{\text{maximum laser power} \times \text{maximum repetition rate}}{\text{actual repetition rate} \times \text{scan speed} \times \text{line spacing}}. \tag{6.5}$$

The relationship between PL intensity and ξ in BSi is also explored. As shown in figure 6.6(a), the PL intensity fluctuates with changes in repetition rate and scan speed, reaching a maximum at $\xi = 52$ kHz, 1 mm s^{-1}.

Following hydrofluoric acid (HF) treatment, the PL signal disappears, confirming that it originates from the oxygen-rich surface layer. Figure 6.6(b) further illustrates a clear dependence of PL intensity on ξ. In the low-ξ regime, increasing ξ enhances luminescence by promoting the formation of luminescent surface structures. In contrast, at higher ξ values, PL intensity becomes more variable—likely due to morphological modifications that alter both the absorptance and the spatial distribution of luminescent features. These findings highlight the importance of tuning ξ to fabricate high-efficiency luminescent structures. In particular, elevated ξ values are favorable for producing BSi with enhanced PL properties, which could be leveraged for advanced sensing applications [54].

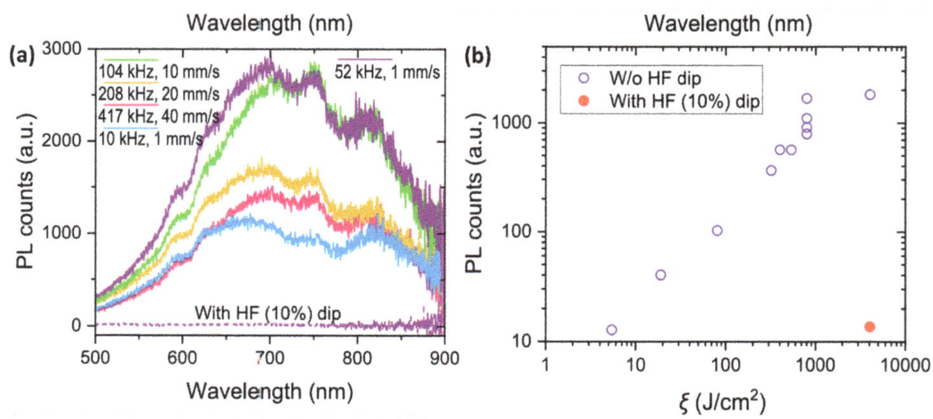

Figure 6.6. (a) PL spectra of BSi samples fabricated with varying repetition rate and scan speed. All samples were characterized immediately after laser processing, except for one reference sample that underwent a 5 min HF (10%) dip. (b) PL intensity dependency (averaged from 500 to 900 nm) on the parameter ξ. (Reproduced with permission from [50]. Copyright 2023 Optica Publishing Group.)

6.6 Summary

This chapter delves into the electronic properties of BSi, emphasizing the influence of its nanostructured morphology on carrier transport, electrical conductivity, carrier lifetime, and recombination dynamics. It also investigates the band structure and density of states, highlighting their significance in enhancing the device performance and broadening the application potential of BSi.

References

[1] Liu X *et al* 2014 Black silicon: fabrication methods, properties and solar energy applications *Energy Environ. Sci.* **7** 3223–63
[2] Liu K *et al* 2014 Sulfur-doped black silicon formed by metal-assist chemical etching and ion implanting *Appl. Phys.* A **114** 765–8
[3] Ravindra N M, Marthi S R and Sekhri S 2016 *Black Silicon: Processing, Properties, and Applications* (New York: Momentum)
[4] Jaeger R C 1987 *Introduction to Microelectronic Fabrication* (Boston, MA: Addison-Wesley Longman)
[5] von Gastrow G *et al* 2017 Recombination processes in passivated boron-implanted black silicon emitters *J. Appl. Phys.* **121** 185706
[6] Brendel R 2003 *Thin-Film Crystalline Silicon Solar Cells: Physics and Technology* (New York: Wiley)
[7] Kittel C and McEuen P 2018 *Introduction to Solid State Physics* (New York: Wiley)
[8] Honsberg C and Bowden S 2019 Types of recombination *Photovoltaics Education* https://www.pveducation.org/pvcdrom/pn-junctions/types-of-recombination
[9] McConnell A and Goodson K E 2005 Thermal conduction in silicon micro- and nano-structures *Annu. Rev. Heat. Transfer* **14** 129–68
[10] Rauh F *et al* 2024 Nanostructured black silicon as a stable and surface-sensitive platform for time-resolved *in situ* electrochemical infrared absorption spectroscopy *ACS Appl. Mater. Interfaces* **16** 6653–64
[11] Oh J, Yuan H-C and Branz H M 2012 An 18.2%-efficient black-silicon solar cell achieved through control of carrier recombination in nanostructures *Nat. Nanotechnol.* **7** 743–8
[12] Liu Y *et al* 2012 Nanostructure formation and passivation of large-area black silicon for solar cell applications *Small* **8** 1392–7
[13] Repo P *et al* 2012 Effective passivation of black silicon surfaces by atomic layer deposition *IEEE J. Photovolt.* **3** 90–4
[14] Winkler M T *et al* 2011 Insulator-to-metal transition in sulfur-doped silicon *Phys. Rev. Lett.* **106** 178701
[15] Fan Z *et al* 2021 Recent progress of black silicon: from fabrications to applications *Nanomaterials* **11** 41
[16] Lin G-R, Lin C-J and Lin C-K 2007 Enhanced Fowler–Nordheim tunneling effect in nanocrystallite Si based LED with interfacial Si nano-pyramids *Opt. Express* **15** 2555–63
[17] Callister Jr W D and Rethwisch D G 2020 *Materials Science and Engineering: an Introduction* (New York: Wiley)
[18] Otto M *et al* 2012 Extremely low surface recombination velocities in black silicon passivated by atomic layer deposition *Appl. Phys. Lett.* **100** 191603

[19] Zaidi S H, Ruby D S and Gee J M 2001 Characterization of random reactive ion etched-textured silicon solar cells *IEEE Trans. Electron Devices* **48** 1200–6

[20] Füchsel K *et al* 2011 Transparent conductive oxides for nano-SIS solar cells *Proc. SPIE* **8065** 80650B

[21] Yoo J, Yu G and Yi J 2009 Black surface structures for crystalline silicon solar cells *Mater. Sci. Eng.* B **159–160** 333–7

[22] Zhao J *et al* 1996 24% efficient perl silicon solar cell: recent improvements in high efficiency silicon cell research *Sol. Energy Mater. Sol. Cells* **41–42** 87–99

[23] Sze S M, Li Y and Ng K K 2021 *Physics of Semiconductor Devices* (New York: Wiley)

[24] Brody J, Rohatgi A and Ristow A 2003 Review and comparison of equations relating bulk lifetime and surface recombination velocity to effective lifetime measured under flash lamp illumination *Sol. Energy Mater. Sol. Cells* **77** 293–301

[25] Hoex B *et al* 2006 Ultralow surface recombination of c-Si substrates passivated by plasma-assisted atomic layer deposited Al_2O_3 *Appl. Phys. Lett.* **89** 042112

[26] Iandolo B *et al* 2019 Black silicon with ultra-low surface recombination velocity fabricated by inductively coupled power plasma *Phys. Status Solidi—Rapid Res. Lett.* **13** 1800477

[27] Penaud J *et al* 2011 Impact of ageing on AlOx layer passivation properties for advanced cell architectures *37th IEEE Photovoltaic Specialists Conference* (Piscataway, NJ: IEEE)

[28] Sperber D, Herguth A and Hahn G 2016 Instability of dielectric surface passivation quality at elevated temperature and illumination *Energy Procedia* **92** 211–7

[29] Niewelt T *et al* 2017 Long-term stability of aluminum oxide based surface passivation schemes under illumination at elevated temperatures *IEEE J. Photovolt.* **7** 1197–202

[30] Otto M *et al* 2015 Black silicon photovoltaics *Adv. Opt. Mater.* **3** 147–64

[31] Savin H *et al* 2015 Black silicon solar cells with interdigitated back-contacts achieve 22.1% efficiency *Nat. Nanotechnol.* **10** 624–8

[32] Allen T *et al* 2014 Reactive ion etched black silicon texturing: a comparative study *IEEE 40th Photovoltaic Specialist Conference (PVSC)* (Piscataway, NJ: IEEE)

[33] Ashcroft N W and Mermin N D 2011 *Solid State Physics* (Boston, MA: Cengage Learning)

[34] Ganeshalingam S, Ford M and Softley T 2013 Ionization of Rydberg H_2 molecules at doped silicon surfaces *J. Chem. Phys.* **138** 114308

[35] Trappmann T, Sürgers C and Löhneysen H v 1997 Observation of P donors on the Si(111) surface by scanning tunneling microscopy *Europhys. Lett.* **38** 177

[36] Trappmann T, Sürgers C and Löhneysen H v 1999 Investigation of the (111) surface of P-doped Si by scanning tunneling microscopy *Appl. Phys.* A **68** 167–72

[37] Jiang H and Chen C 2015 Analysis and calculation of electronic properties and light absorption of defective Sulfur-doped silicon and theoretical photoelectric conversion efficiency *J. Phys. Chem.* A **119** 3753–61

[38] Liu C *et al* 2021 High-efficiency black silicon tunnel oxide passivated contact solar cells achieved by adjusting the boron diffusion process *J. Mater. Sci., Mater. Electron.* **32** 23465–71

[39] Padrez Y and Golubewa L 2024 Black silicon surface-enhanced Raman spectroscopy biosensors: current advances and prospects *Biosensors* **14** 453

[40] Aleem M, Vishnuraj R and Pullithadathil B 2025 Recycled silicon solar cell-derived nanostructured p-black silicon device for high performance NO_2 gas sensor applications *RSC Appl. Interfaces* **2** 220–9

[41] Hartley J S *et al* 2015 Black silicon as a platform for bacterial detection *Biomicrofluidics* **9** 061101

[42] Jin X *et al* 2021 High-performance black silicon photodetectors operating over a wide temperature range *Opt. Mater.* **113** 110874
[43] Garin M *et al* 2020 Black-silicon ultraviolet photodiodes achieve external quantum efficiency above 130% *Phys. Rev. Lett.* **125** 117702
[44] Golubewa L *et al* 2023 Stable and reusable lace-like black silicon nanostructures coated with nanometer-thick gold films for SERS-based sensing *ACS Appl. Nano Mater.* **6** 4770–81
[45] Tan Q *et al* 2019 Nano-fabrication methods and novel applications of black silicon *Sens. Actuators* A **295** 560–73
[46] Lee J *et al* 2017 Biocompatible multifunctional black-silicon for implantable intraocular sensor *Adv. Healthcare Mater.* **6** 1601356
[47] Lv J *et al* 2018 Review application of nanostructured black silicon *Nanoscale Res. Lett.* **13** 110
[48] Alsolami A *et al* 2024 Recent advances in black silicon surface modification for enhanced light trapping in photodetectors *Appl. Sci.* **14** 9841
[49] Zhao Z *et al* 2023 Black silicon for near-infrared and ultraviolet photodetection: a review *APL Mater.* **11** 021107
[50] Radfar B *et al* 2023 Optoelectronic properties of black silicon fabricated by femtosecond laser in ambient air: exploring a large parameter space *Opt. Lett.* **48** 1224–7
[51] Nava G *et al* 2013 Scaling of black silicon processing time by high repetition rate femtosecond lasers *Opt. Mater. Express* **3** 612–23
[52] Chen T *et al* 2011 Luminescence of black silicon fabricated by high-repetition rate femtosecond laser pulses *J. Appl. Phys.* **110** 073106
[53] Emelyanov A V *et al* 2012 Visible luminescence from hydrogenated amorphous silicon modified by femtosecond laser radiation *Appl. Phys. Lett.* **101** 081902
[54] Mironenko A Y *et al* 2019 Ultratrace nitroaromatic vapor detection via surface-enhanced fluorescence on carbazole-terminated black silicon *ACS Sens.* **4** 2879–84

Chapter 7

Optical properties

This chapter explores the optical properties of black silicon (BSi), focusing on light trapping, antireflective characteristics, and surface morphology. It covers various optical simulation methods, including effective medium approximation, rigorous coupled-wave analysis, the finite-difference time-domain, and angular distribution of the cone model, which help to determine the optical constants of BSi. Photoluminescence in BSi is briefly discussed. The chapter also highlights the applications of BSi in photovoltaics, photodetectors, and biosensors, with particular attention on the fabrication of biosensors based on surface-enhanced Raman spectroscopy (SERS).

7.1 Introduction

Black silicon (BSi) is a nanostructured form of crystalline silicon (c-Si) with enhanced optical properties, making it valuable in optoelectronics, photovoltaics (PVs), and sensors [1, 2]. Its surface nanostructures significantly improve light absorption and reduce reflection, lowering reflectance from 30%–40% (in c-Si) to less than 5% across a broad spectral range [3]. This effect is achieved through enhanced light scattering and an extended optical path, increasing absorption in the ultraviolet (UV), visible, and near-infrared (NIR) spectra. BSi has a variable refractive index (n) due to its textured surface and porosity, unlike c-Si, which has a fixed refractive index (~3.5 at 600 nm). This gradient refractive index reduces reflection losses. Additionally, its higher extinction coefficient (k) enhances light absorption, improving its efficiency in optoelectronic applications. The rough surface of BSi enhances photon trapping through Rayleigh and Mie scattering [4], improving light absorption and boosting power conversion efficiency, especially in thin-film solar cells. Its strong NIR and mid-infrared (mid-IR) absorption also enhances IR sensing performance [5]. Surpassing crystalline silicon in absorption, reflection, and scattering, BSi is ideal for high-efficiency photovoltaics, photodetectors, and infrared devices [1]. However, surface recombination losses require effective passivation.

doi:10.1088/978-0-7503-3247-7ch7

Table 7.1. Comparison of the properties of BSi with c-Si.

Property	Black silicon (BSi)	Crystalline silicon (c-Si)
Absorption [7]	High (broadband, including IR)	Moderate (mostly visible)
Reflectance [3]	Near-zero (~1%–5%)	High (~30%–40%)
Refractive index [8]	Gradient-index (variable)	Fixed (~3.5 at 600 nm)
Extinction coefficient [8]	High (strong light absorption)	Low
Scattering [4]	Strong (Rayleigh and Mie effects)	Weak (without texturing)

With significant advancements in fabrication and passivation methods, their broader commercial adoption has been facile [6]. Table 7.1 compares some important properties of BSi with those of c-Si.

7.2 Fundamentals of optical properties

The optical properties of a material define its interaction with light, including the associated optical phenomena such as absorption, reflection, transmission, refraction, and scattering. These characteristics are essential in areas of applications such as photovoltaics, optics, and material science [1, 2]. These interactions are governed by parameters such as the refractive index, extinction coefficient, and dielectric function. The following section presents important equations describing the optical behavior of materials.

Figure 7.1(A) illustrates a simple schematic diagram representing the reflection and transmission of an incident beam. Refractive index (n) determines how light propagates through the material, i.e. n measures how much light slows down in a material. It is defined as the ratio of the speed of light in a vacuum (c) to the speed of light in the material (v):

$$n = \frac{c}{v}. \tag{7.1}$$

For materials with absorption properties, the refractive index is complex and is defined as

$$\tilde{n} = n + \mathrm{i}k, \tag{7.2}$$

where n is the real part of the equation and is related to phase velocity, and k is the imaginary part, called the extinction coefficient, and is related to absorption. Materials with higher refractive indices bend light more (e.g. silicon: $n \approx 3.5$ at 600 nm) [9].

The absorption coefficient (α) is characteristic of a particular material; it quantifies the absorbed light as it travels through a material. It is related to the extinction coefficient and wavelength (λ) [10]. It follows the Beer–Lambert law and is written as

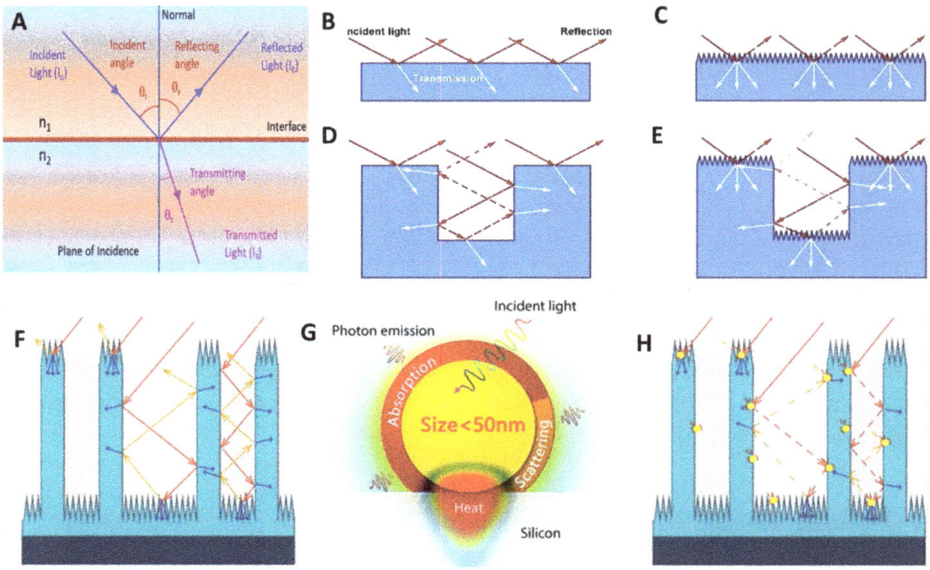

Figure 7.1. (A) Schematic diagram exhibiting the interaction of an incident beam with a material interface, including reflection and transmission. θ_i, θ_r, and θ_t are the angles made by incident, reflected, and transmitted rays to the normal at the interface, respectively. Schematic illustration of the antireflection principle of optical waves interacting with surfaces of varying structural features: (B) flat surface, (C) surface with nanopores, (D) surface with micropores, and (E) surface with combined micro-nanopores. (F) Light-trapping mechanism in BSi with a hierarchical texture. (G) Schematic illustration highlighting dominant light absorption in Au NPs smaller than 50 nm and (H) antireflection effect resulting from the synergy between Au NP-induced localized surface plasmon resonance (LSPR) and the hierarchical texture. ((B), (C), (D), and (E) reproduced with permission from [16]. Copyright 2019 Elsevier. (F), (G), and (H) reproduced with permission from [17]. Copyright 2020 Wiley, CC BY 4.0.)

$$\alpha = \frac{4\pi k}{\lambda}. \tag{7.3}$$

The intensity of light (I) decays exponentially with distance (x) in an absorbing material and is given by

$$I(x) = I_o e^{-\alpha x}, \tag{7.4}$$

where I is the intensity of the beam coming out of the material, I_o is the intensity of the incident beam (figure 7.1(A)), and x is the path through which the photons traverse. Higher values of α mean stronger absorption (e.g. metals have high absorption, while transparent materials have low absorption). High k values indicate strong absorption, as seen in metals such as gold and silver [11].

Reflectivity (R) is a fraction of incident light at a material interface (figure 7.1(A)). Fresnel's equations predict reflection at a planar interface by the following relation:

$$R = \left| \frac{n_1 \cos\theta_i - n_2 \cos\theta_t}{n_1 \cos\theta_i + n_2 \cos\theta_t} \right|^2, \tag{7.5}$$

where n_1 and n_2 are the refractive indices of two media [12] (figure 7.1(A)) and θ_i, θ_r, and θ_t are the angles that the incident, reflected, and transmitted rays make with respect to the normal, respectively. The value of R depends on the angle of incidence. For the case of normal incidence, $\theta_i = \theta_t = 0$; then equation (7.5) reduces to

$$R = \left(\frac{n_1 - n_2}{n_1 + n_2}\right)^2. \tag{7.6}$$

For second medium having extinction coefficient k, R can be calculated as

$$R = \left[\frac{(n_1 - n_2)^2 + k^2}{(n_1 + n_2)^2 + k^2}\right]. \tag{7.7}$$

In BSi, the graded refractive index reduces R by creating a gradual transition from air to silicon.

Transmittance (T) is defined as a fraction of light beam that is not reflected or absorbed but is transmitted through the material. The optical property T is important in optical coatings and photovoltaic materials [13] and depends on the absorption coefficient. The transmittance for a material of thickness t is given by

$$T = \mathrm{e}^{-\alpha t}. \tag{7.8}$$

Absorptance is the fraction of light that is absorbed into the material and is defined as

$$A(\lambda) = 1 - R(\lambda) - T(\lambda), \tag{7.9}$$

for non-absorbing material

$$T(\lambda) = 1 - R(\lambda). \tag{7.10}$$

The dielectric function (ε) describes how a material responds to an electric field [14]. It is related to the complex refractive index and is written as

$$\tilde{\varepsilon} = \varepsilon_1 + \mathrm{i}\varepsilon_2 = \tilde{n}^2 = (n + \mathrm{i}k)^2. \tag{7.11}$$

Expanding equation (7.11) gives

$$\varepsilon_1 = n^2 - k^2 \tag{7.12}$$

$$\varepsilon_2 = 2nk, \tag{7.13}$$

where ε_1 and ε_2 represent the real part (related to polarization) and imaginary part (related to energy absorption) in equations (7.12) and (7.13), respectively.

In BSi, because of its roughness, light is scattered from its surface. Scattering occurs when light is deflected (or redirected) by inhomogeneities (or surface roughness) in the material. Rayleigh scattering occurs for small particles (λ-dependent) and Mie scattering occurs for larger particles [15].

The scattering cross-section (σ_s) quantifies the efficiency of scattering and is defined as

$$\sigma_s = \frac{I_{scattered}}{I_{incident}}. \tag{7.14}$$

The intensity of scattered light (I_s) depends on the wavelength (λ) and particle size (d) and is defined by the following relation:

$$I_s \propto \frac{d^6}{\lambda^4}. \tag{7.15}$$

Nanostructures smaller than the wavelength of light act as a broadband antireflection coating, suppressing reflection across a wide spectrum. The light-trapping efficiency is often compared to the Lambertian limit, where the path length enhancement factor G is maximized and is given by

$$G = \frac{4n^2}{\sin^2\theta_i}. \tag{7.16}$$

Nanostructures induce diffuse scattering, randomizing light paths and approximating a Lambertian surface, which increases internal incidence angles and enhances absorption toward the theoretical limit.

7.3 Influence of structure on optical properties

The specific geometry and arrangement of nanostructures in BSi significantly influence its optical properties. Features such as height, diameter, and spacing of nanostructures affect the extent of light scattering and absorption. Additionally, the periodicity of these structures can lead to photonic crystal effects, further optimizing light management and minimizing reflection losses [16]. The following section will outline the light-trapping capabilities, including antireflective properties and surface morphology, of BSi.

7.3.1 Light trapping and antireflective properties

BSi maintains low reflectance over a wide range of angles of incidence, unlike traditional antireflective coatings. The antireflection (AR) principle of optical waves is illustrated in figure 7.1(B) which depicts the reflection behavior on a planar surface [16]. Figure 7.1(C) shows that the nanopores on the surface, with dimensions comparable to optical wavelengths, significantly alter reflectivity through multiple refractions and transmissions. Figure 7.1(D) illustrates that the micropores, which are larger than the wavelength of incident light, facilitate continuous reflection along the depth direction due to their vertical smooth sidewalls [16]. These micron-scale pores enhance photon absorption primarily through geometric optics effects, allowing some reflected light to re-enter the silicon, thereby extending the optical path length and improving light–material coupling. Figure 7.1(E) [16] demonstrates that the nanopores at the bottom absorb weakened light (intensity) and adjust the tilt angle of the reflected portion, further increasing the optical path length. Eventually, this ordered–disordered multiscale pore structure enhances light absorption, leading to improved optical performance [16].

In another interesting study, BSi with novel chimney-like hierarchical micro/nanostructures is created using two-step reactive ion etching. Gold nanoparticles (Au NPs) are then deposited onto the BSi via magnetron sputtering. The synergy between the micro/nanostructures and Au NPs effectively suppresses light reflection, achieving an ultralow broadband reflection of less than 1% across the wavelength range of 220–2600 nm [17]. Figure 7.1(F) demonstrates the light-trapping mechanism in BSi with a hierarchical texture. This hybrid texture design effectively reduces specular reflection losses, enhances light–material coupling, and extends the optical path length. This combination leads to ultralow reflectance for wavelengths below 1100 nm. Figure 7.1(G) illustrates the absorption-dominant behavior of Au NPs with sizes smaller than 50 nm, highlighting the transfer of thermal energy and hot carriers at the interface between the localized surface plasmon resonance (LSPR) active particles and bulk silicon [17]. This investigation has been motivated to explore the optimal configuration of Au NPs with LSPR on BSi to further reduce reflectivity above 1100 nm through thermal-dissipation-dominant energy decay. Figure 7.1(H) illustrates the potential mechanism of the synergistic trapping of incident light. Light striking the top surface is less likely to undergo specular reflection due to photon absorption and scattering induced by metallic particles. The light incident on the sidewalls is attenuated through plasmon oscillations in the NPs and absorption within the silicon. Any reflected or scattered light may be further weakened upon interacting with metallic particles on the opposite sidewalls. Additionally, nanopores and NPs at the bottom of the structure can capture the residual reflected and scattered light. Based on this analysis, achieving broadband ultralow reflectivity is a promising prospect. A systematic investigation into the synergy between hierarchical textures and Au NP-induced LSPR is essential to enhance the antireflective properties of BSi [17].

7.3.2 Surface morphology

The surface morphology of BSi, with nanostructures such as spikes, cones, and pores, significantly enhances light trapping and optical absorption by reducing reflection and increasing internal scattering. High-aspect-ratio structures boost broadband absorption, and are particularly beneficial for thin-film silicon devices. However, taller structures may increase surface recombination losses, which can be mitigated through passivation with dielectric layers such as silicon dioxide or aluminum oxide. Various fabrication methods, such as femtosecond laser processing, can optimize absorption, especially in the NIR region. Overall, the textured surface extends the optical path length, improving absorption efficiency in thinner silicon layers.

Figure 7.2(a) presents images of inductively coupled plasma (ICP) textured as-cut multicrystalline (mc-Si) wafers etched for 1–4 min. As the etching time increases, wafer reflection decreases, resulting in a black appearance [18]. Additionally, it has been found that the ICP process is independent of crystal orientation and surface roughness, enabling the texturing of monocrystalline, multicrystalline, and kerf-less wafers using the same set of parameters [18]. Figure 7.2(b) presents tilted-view

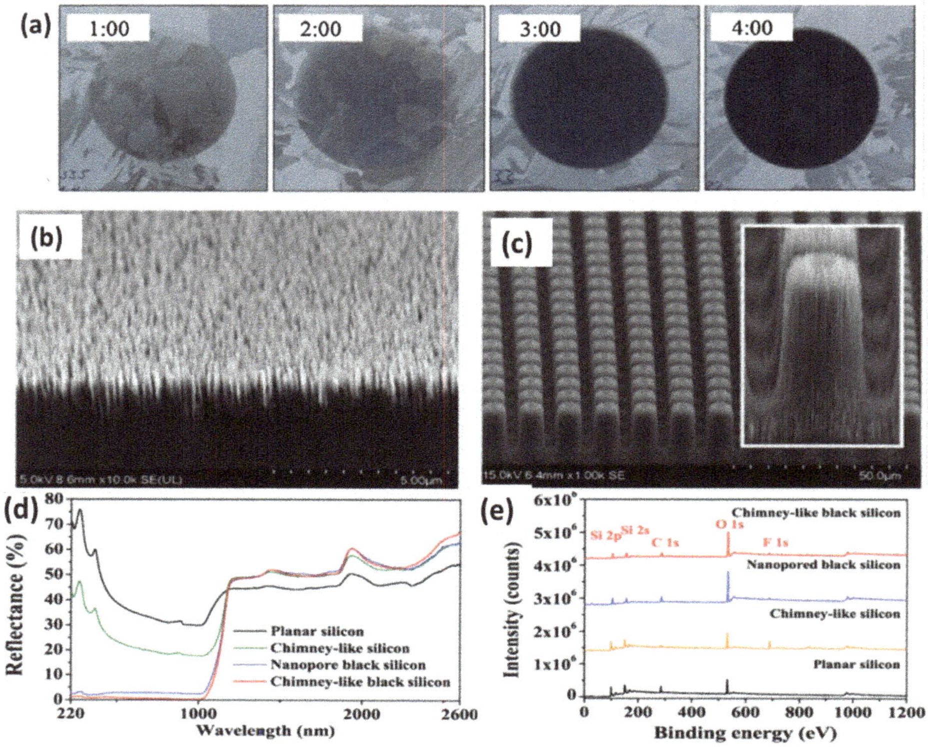

Figure 7.2. (a) Photographs of as-cut multicrystalline silicon wafers after texturing (textured diameter = 5 cm) via ICP for durations (1–4 min), illustrating changes in reflectivity. Characterization of BSi with various structural features: (b) SEM images of BSi with nanopores in a tilted view. (c) A tilted view of chimney-like BSi with a hierarchical structure; the inset shows a magnified chimney decorated with nanopores. (d) Reflection spectra of polished and textured silicon surfaces. (e) X-ray photoelectron spectra of polished and textured silicon surfaces. ((a) Reproduced with permission from [18]. Copyright 2016 Elsevier. (b), (c), (d), and (e) Reproduced with permission from [17]. Copyright 2020 Wiley, CC BY 4.0.)

scanning electron microscopy (SEM) images of the silicon surface textured using maskless reactive ion etching (RIE) with an SF_6/O_2 gas mixture. The resulting nanopores exhibit an average depth of 6.5 μm and a diameter of less than 450 nm [17]. By increasing the SF_6/O_2 gas flow ratio, the surface texture can be modified from microporous to nanoporous structures [19]. Figure 7.2(c) shows a tilted-view SEM image of BSi featuring nanopores located on both the top and bottom of the chimney-like arrays. Figure 7.2(d) illustrates the reflectance spectra, revealing that silicon samples with micro- and nanostructures exhibit significantly lower reflectance compared to planar silicon in the visible range. Among them, nanopore BSi demonstrates the lowest reflectance. Additionally, BSi with hierarchical structures (chimney-like BSi) exhibits even lower reflectance than nanopore-only BSi within the 220–1100 nm wavelength range, while both display similar reflectance behavior at wavelengths above 1100 nm [17]. X-ray photoelectron spectroscopy (XPS) analysis, as shown in figure 7.2(e), was conducted to examine the elemental composition of both planar and textured silicon surfaces. The results reveal that silicon, carbon, oxygen, and fluorine are the primary elements present on these surfaces [17].

7.4 Computational methods for simulation of optical properties

The optical properties of BSi have been analysed using computational modeling to predict its performance in device applications such as in photovoltaics and photodetectors. These simulations help in understanding the light interactions with nanostructured surfaces, including absorption, reflection, and scattering. The optical modeling of BSi focuses on its enhanced light trapping and absorption due to its nanostructured surface, which minimizes reflectance. Chapter 3 briefly discusses the simulation methods; this section offers a more detailed exploration of selected methods.

7.4.1 Effective medium approximation

The effective medium approximation (EMA) is commonly used to model BSi as a graded-index medium, simplifying complex surface structures into a homogenized layer [8, 9, 20–22]. By representing BSi as a multi-layer structure with varying refractive indices, optical properties such as the refractive index and extinction coefficient can be predicted. In a related study, the EMA is applied to model BSi nanocones and nanowires to predict their optical behavior, including absorption and reflection characteristics [22].

The EMA [21, 23] is based on an assumption that an inhomogeneous material is composed of two types of spherical grains with complex-valued permittivity, ε_1 and ε_2, mixed in volume fractions of (f) and ($1-f$) [20, 24], where f is the volume filling factor. The derivation of effective permittivity ε^* is based on the assumption that the material surrounding the inclusions can be replaced by a homogeneous matrix with the same effective permittivity ε^* (f, ε_1, ε_2). First order approximation is used for a dilute composite to calculate the resulting effective property. For simulation, a vertically aligned silicon nanowire array of thickness H_w is considered on a semi-infinite silicon substrate, as shown in figure 7.3(a). Each nanowire has a diameter d and an average spacing a, with air as the surrounding medium. The volume filling ratio f is given by $f = \pi d^2(4a^2)$. The silicon substrate is assumed to be opaque, as the penetration depth of silicon is typically small, except near the indirect bandgap ($\sim$1.1 μm). The model assumes that the geometric dimensions are much smaller than the wavelength of electromagnetic waves. Based on Bruggeman's effective medium approximation, the effective dielectric function ε_{eff} of the BSi layer can be calculated by solving the following equation:

$$\frac{(1-f)(\varepsilon_{air}-\varepsilon_{eff})}{\varepsilon_{eff}+g(\varepsilon_{air}-\varepsilon_{eff})}+\frac{f(\varepsilon_{Si}-\varepsilon_{eff})}{\varepsilon_{eff}+g(\varepsilon_{Si}-\varepsilon_{eff})}=0, \tag{7.17}$$

where ε_{air} and ε_{Si} are the dielectric functions of air and Si, respectively. The depolarization factor g depends on the geometry and polarization. Figure 7.3(b) presents an Illustration of the wave propagation in a three-layer structure used in the

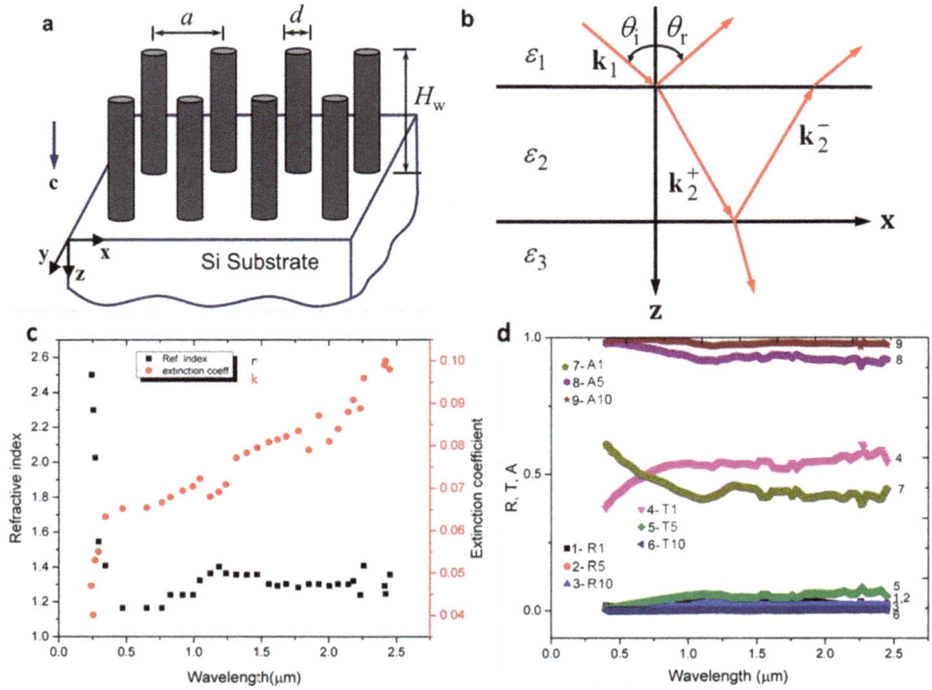

Figure 7.3. (a) Schematic of a silicon nanowire array on a silicon substrate, where 'c' represents the optical axis. (b) Illustration of wave propagation in a three-layer structure used in the model, with medium 2 being a uniaxial medium, where the optical axis is aligned with the z-axis. (c) Effective refractive index (n_{eff}) and the effective extinction coefficient (k_{eff}) for BSi as a function of wavelength. (d) The simulated reflectance (R), transmittance (T), and absorptance (A) of BSi for various thicknesses of BSi (1, 5, and 10 μm) as a function of wavelength. ((a) and (b) Reproduced with permission from [20]. Copyright 2013 Elsevier Masson SAS. (c) and (d) Reproduced with permission from [8]. Copyright 2015 Springer.)

model, with medium 2 being a uniaxial medium, where the optical axis is aligned with the z-axis.

7.4.1.1 Optical constants of black silicon

The reported work in [8] uses the effective medium approximation to calculate the optical constants, refractive index and extinction coefficient. Since BSi is a surface modification of silicon rather than a distinct material, defining its exact optical constants is challenging [25]. The refractive index and extinction coefficient of BSi are not fixed but depend on its morphology [21]. The effective medium approximation [9, 21] can be used to determine the optical constants of BSi. The resulting optical constants include the effective refractive index (n_{eff}) and the effective extinction coefficient (k_{eff}). The effective n and k are determined using the R, T, and A values from equations (7.6), (7.8), and (7.9) [8]. Figure 7.3(c) illustrates the optical constants n_{eff} and k_{eff}, determined using EMA. A sharp change in k is observed between 1.1 and 1.2 μm, corresponding to the energy bandgap of crystalline silicon. Additionally, figure 7.3(d) presents A, R, and T as functions of wavelength for BSi layers of different thicknesses (1, 5, and 10 μm). Absorptance increases with thickness, while reflectance and transmittance decrease, demonstrating the enhanced light-trapping properties of thicker BSi layers.

7.4.2 Rigorous coupled-wave analysis

Rigorous coupled-wave analysis (RCWA) is a powerful technique for modeling the optical properties of BSi, especially for periodic nanostructures [26–29]. This method solves Maxwell's equations by expanding electromagnetic fields into Fourier series, making it highly effective in analysing diffraction efficiencies, light absorption, and reflection. RCWA is helpful in analysing the impact of periodic surface features on the optical performance of BSi, aiding in the design of high-efficiency optoelectronic devices [30].

As shown in figure 7.4(a), for BSi simulation, RCWA defines three regions: (i) homogeneous incident medium (region I) with a refractive index of 1 (air);

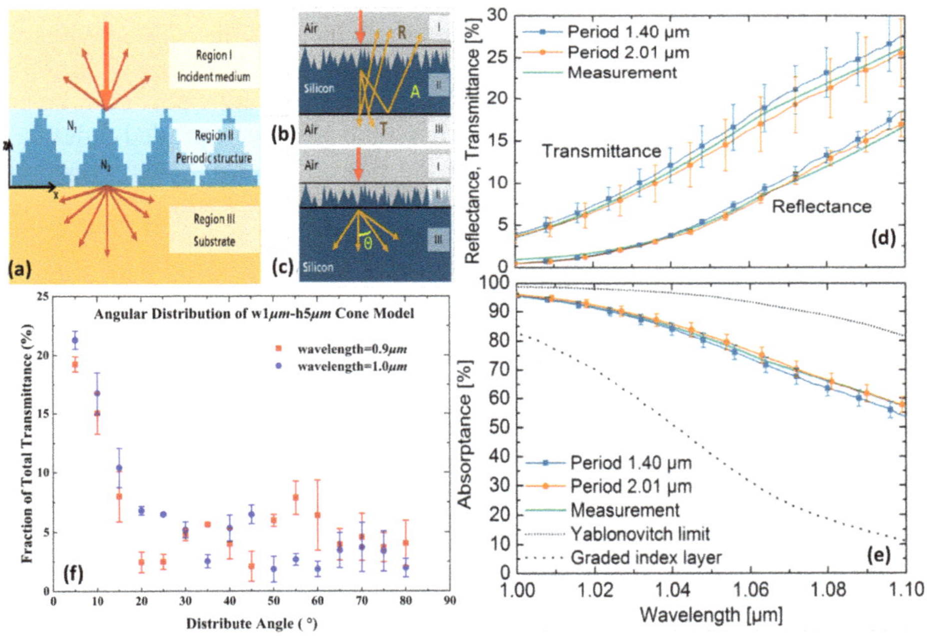

Figure 7.4. (a) A schematic illustration of the three regions necessary for the RCWA simulation. The medium of incidence (region I) and the substrate (region III) are homogeneous, while the medium in between is a periodic structure (region II). The z-direction is discretized in a definite number of layers which are periodic in the x- and y-directions. (b) Simulation of reflectance (R), transmittance (T), and absorptance (A), silicon is treated as a thick layer within region II. (c) Simulation of the angular distribution of light within the silicon bulk; the substrate (region III) is also modeled as silicon. (d) The simulated reflectance and transmittance of a 250 μm thick silicon wafer with BSi on the front side, demonstrating strong agreement with measured data. (e) Comparison of simulated and measured absorptance with that of a graded-index layer and an ideal diffuse scatterer (approaching the Yablonovitch limit). (f) Angular distributions of the w1–h5 cone model at wavelengths of 0.9–1.0 μm. ((a), (b), (c), (d), and (e) Reproduced with permission from [26]. Copyright 2016 Optica Publishing Group. (f) Reproduced from [31]. Copyright 2018 AIP Publishing, CC BY 4.0.)

(ii) periodic structure (region II), consisting of 160 layers (each 6.25 nm thick, 1 μm total height), with refractive index variation in the z-direction and periodic functions in the x- and y-directions, optimized through convergence analysis [26]; and (iii) homogeneous medium substrate (region III). This multi-layer configuration enables precise modeling of light interactions in BSi, accurately capturing reflection and transmission phenomena. For RCWA simulation of reflection and transmission, air is used as the substrate medium, with a 250 μm thick silicon layer in region II (figure 7.4(b)). Due to interference effects in the silicon bulk layer, the RCWA simulation exhibits strong oscillations in reflectance and transmittance as it progresses coherently. Therefore, reflectance and transmittance are computed with a 1 nm wavelength increment and then averaged over 20 neighboring points at each wavelength. For the angular light distribution in silicon, the substrate is modeled as non-absorbing silicon ($k = 0$), as shown in figure 7.4(c) [26].

Figure 7.4(d) compares simulated reflectance, transmittance, and absorptance (figure 7.4(e)) spectra for BSi structures with two unit cell sizes: 1.40 μm (averaged over 15 random structures with standard deviation error bars) and 2.01 μm (from 10 random structures). For comparison, the graph includes the Yablonovitch limit (ideal absorptance with perfect ARC at the front side and Lambertian scattering at the rear) [32] and the absorptance of a silicon wafer with a graded-index layer on the front. The absorptance of BSi is higher than that of a graded-index layer, indicating light path enhancement due to the BSi on the front side. However, it remains lower than the Lambertian scatterer, suggesting the potential for further improvement through enhanced scattering properties or rear-side light-trapping structures. Additionally, the simulated spectra in figures 7.4(d) and (e) match the measured data well for these unit cell sizes, although this may not apply universally to all sizes [26].

7.4.3 Finite-difference time-domain

Finite-difference time-domain (FDTD) simulations are used to model light interaction with the nanostructured surface of BSi. FDTD solves Maxwell's equations on a discrete grid to analyse light propagation, absorption, and scattering in complex structures [31, 33–35]. A study using FDTD simulations examines the impact of surface features on light absorption, offering valuable insights for designing BSi surfaces to enhance optical performance [31] and investigates the BSi surface using two models, rough surface and cone models, via FDTD simulation [31]. The following section will analyse the angular distribution of the cone model for BSi.

7.4.3.1 Angular distribution of the cone model

The optical properties of BSi are influenced by the angular orientation and morphology of its surface. The height and alignment of the nanostructures play a key role in light scattering and absorption. A study investigates the impact of nanostructure height on reflectance and absorption, providing valuable insights for enhancing its optical characteristics [31]. In the study, the simulation focuses on the w1–h5 cone model, where the cone width is 1 μm and the height of the cone is 5 μm.

Table 7.2. Comparison table for EMA, RCWA, and FDTD methods.

Feature	EMA [8, 9, 20–22]	RCWA [26–29]	FDTD [31, 33–35]
Method type	Analytical approximation	Frequency-domain numerical method	Time-domain numerical method
Geometry suitability	Homogeneous or layered materials	Periodic nanostructures (e.g. gratings)	Complex, non-periodic geometries
Computational cost	Low	Moderate	High
Output domain	Effective optical properties	Reflectance, transmittance, diffraction spectra	Broadband response, transient effects
Accuracy	Low for complex structures	High for periodic structures	Very high for complex and irregular structures
Broadband analysis	Limited	Multiple runs needed	Single run covers broadband
Surface roughness	Not well suited	Suitable for periodic roughness	Suitable for both periodic and random roughness
Transient effects	Not supported	Not supported	Supported

In order to observe the angular distribution of light over a wide spectral range, the simulation wavelength is set from 0.4 to 3 μm. A single cone spike is simulated to avoid wave interference. Transmission angles are analysed using a ratio-based method, with 5° intervals corresponding to specific monitor lengths. Figure 7.4(f) presents the angular distributions of the w1–h5 cone model at wavelengths of 0.9–1.0 μm, and shows that as the wavelength increases, the transmitted light increases at lower angles and decreases at higher angles [31].

These modeling approaches are crucial for understanding the optical behavior of BSi and optimizing its properties for various applications. They provide valuable insights into how nanostructure geometry affects optical performance, aiding in the design and development of BSi-based devices.

Table 7.2 highlights the strengths and limitations of each aforementioned method depending on the structure, computational requirements, and desired output.

7.5 Photoluminescence in BSi

Photoluminescence (PL) is a powerful technique for characterizing BSi, offering valuable insights into its optoelectronic properties and enhancing its applications in photovoltaics, photodetectors, biosensors, and light-emitting devices [36–38]. PL in BSi originates from modified surface properties, increased defect states, and quantum confinement effects due to nanostructuring. These factors influence charge carrier recombination, enhancing visible and near-infrared PL while slightly altering the bandgap, thus affecting absorption and emission properties [36, 37]. As shown in the schematic diagram (figure 7.5(a)), when charge carriers (for example, electrons)

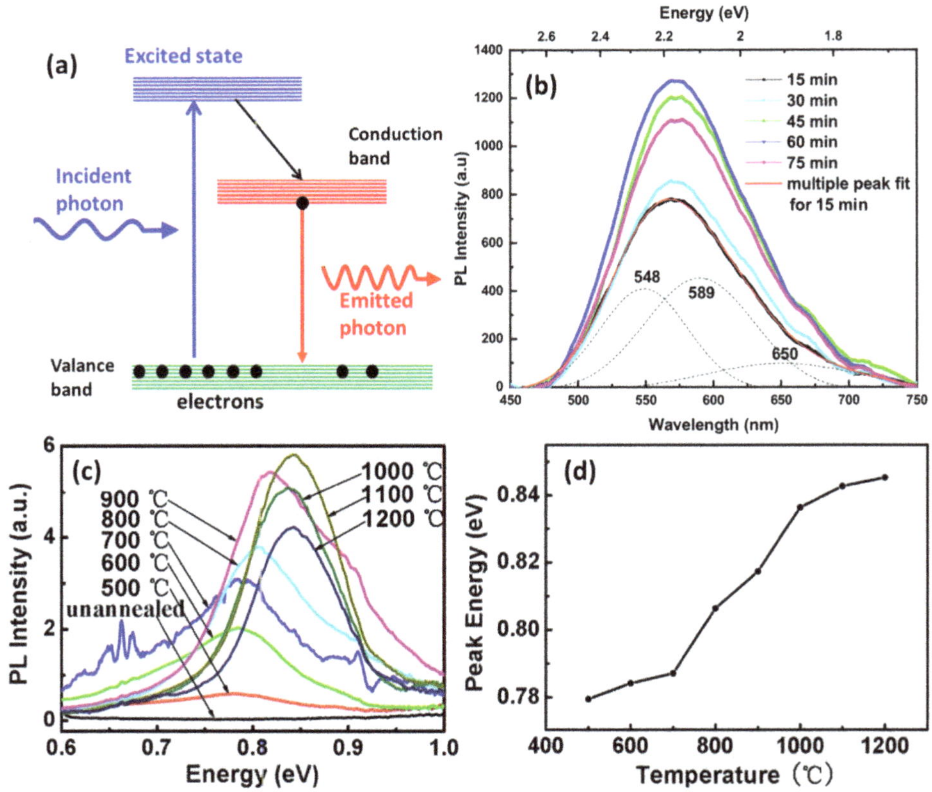

Figure 7.5. (a) Principle of PL spectroscopy. (b) PL spectra of a porous silicon sample for different etching times: 15, 30, 45, 60, and 75 min. A simulated PL peak for the 15 min etching sample is also included for comparison. (c) PL spectra of BSi samples before and after annealing at various temperatures. (d) Dependence of PL peak energy on annealing temperature. ((b) Reproduced with permission from [39]. Copyright 2019 Springer Nature. (c) and (d) Reproduced with permission from [40]. Copyright 2013 Optica Publishing Group.)

in a material are excited by a light source, they absorb photons and transition to a higher energy state. As the excited electrons return to their ground state, they emit photons with characteristic wavelengths. By analysing the emitted light, PL spectroscopy provides insights into bandgap energy, defect states, and recombination mechanisms in semiconductors, nanomaterials, and other optoelectronic materials. The technique is widely used to evaluate surface passivation and defect levels, although high temperatures can degrade PL efficiency by affecting surface passivation. The intensity and wavelength of PL depend on nanostructure size, distribution, doping levels, and surface passivation. The presence of hydrogen or silicon nitride enhances PL by reducing non-radiative recombination. Quantum confinement effects and doping (n-type or p-type) further modify PL characteristics.

In one study [39], upgraded metallurgical grade silicon (UMG-Si) wafers were used to create mesoporous nanostructures as effective antireflection layers for solar photovoltaic cells. The length of vertical silicon nanostructure arrays was controlled by adjusting the etching time during the MACE process with a silver catalyst. Optical, structural, and morphological properties, along with antireflection characteristics, were analysed. Figure 7.5(b) shows the PL spectra of mesoporous silicon for various etching durations. The PL spectra can be deconvulated into three peaks, with centers at approximately 550, 590, and 660 nm. Three main PL peaks were observed, originating from silicon nanocrystals (SiNCs) on the surface and silicon nanowires decorated with SiNCs. Optical analysis showed that the 45 min etched sample exhibited strong absorption in the UV–blue spectral range, whereas all other samples also displayed absorption in the green spectral range [39].

Strong infrared PL was observed for the first time in BSi formed through femtosecond laser irradiation followed by rapid thermal annealing [40]. Figure 7.5 (c) presents the PL spectra of BSi samples, both unannealed and annealed at various temperatures. The measurements were conducted at 15 K with an excitation power of 400 mW. For the unannealed BSi sample, no PL signal was detected in the infrared region, even when the excitation power was increased to 1000 mW. In contrast, a strong PL signal was observed in the sample annealed at 500 °C, indicating the significant effect of annealing on enhancing infrared PL. Figure 7.5(d) illustrates the variation of PL peak energy as a function of annealing temperature. The PL peak energy shifted from 0.78 to 0.84 eV (corresponding to 1.59–1.48 μm) as the annealing temperature increased from 500 °C to 1200 °C. After annealing at higher temperatures (1100 °C and 1200 °C), both the PL peak intensity (not shown here) and peak energy showed minimal changes compared to samples annealed at 1000 °C, indicating high thermal stability of the BSi structures and their PL properties. Therefore, the PL spectra reveal a shift in the PL peak intensity and wavelength as the annealing temperature increases. This shift in peak energy is believed to be associated with the changes in the strain states within the annealed BSi samples. Strain in BSi can alter the energy levels of impurities or defects, and even modify the bandgap, ultimately leading to variations in the observed PL peak energy [40].

7.6 Applications of black silicon

BSi's distinctive optical and electronic properties make it highly versatile across multiple applications. The advancements emphasize continued research to optimize the optical properties of BSi and its derivatives, enhancing performance in applications such as photodetectors, solar cells, and biosensors [2].

7.6.1 Black silicon in photovoltaics

BSi has attracted significant interest in photovoltaic research due to its superior light-absorbing properties, which enhance solar cell efficiency. Recent advancements include the development of 'black silicide', a novel BSi material with better light-trapping capabilities and higher NIR absorption (1100–1800 nm) than traditional BSi. This makes it a strong candidate for tandem PV devices, potentially increasing photogenerated current densities [41]. Additionally, integration of BSi into tandem structures, such as c-Si–BSi–perovskite solar cells, has shown improvements in light absorption and overall efficiency [42].

In one study, an 18.2%-efficient solar cell with BSi nanostructure has been designed that eliminates the need for AR coatings. The fabrication method for the BSi is described in detail in [43]. Figure 7.6(A) displays the illuminated current density–voltage (*J–V*) curve of the high performance nanostructured BSi solar cell, as independently measured by NREL. An SEM image of this nanostructured silicon solar cell is shown in the inset of figure 7.6(A). The solar cell exhibits performance parameters, namely an open circuit voltage (V_{oc}) of ~0.628 V, a short circuit current density (J_{sc}) of 36.45 mA cm^{-2}, a fill factor (FF) of 79.6%, and efficiency = 18.2% under AM 1.5 G illumination. Figure 7.6(A) also compares the results with the *J–V* curves of a polished solar cell with no AR coating and a cell with the industry-standard pyramid texture and AR coating [43].

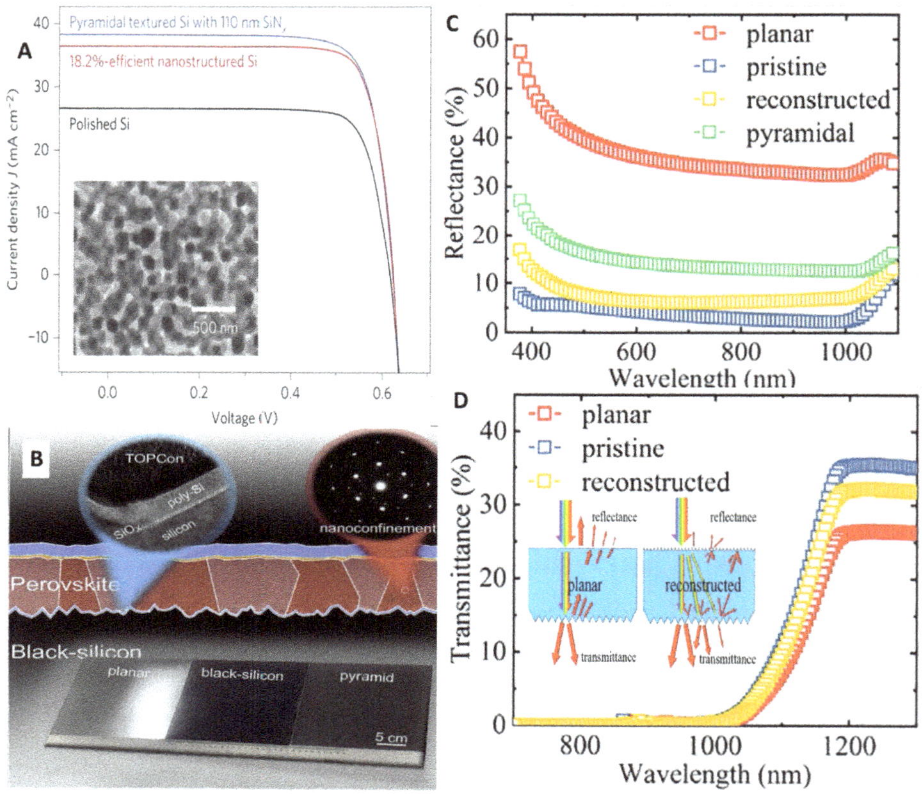

Figure 7.6. (A) Current density–voltage (*J–V*) curves of 18.2%-efficient nanostructured BSi, polished silicon, and pyramid-textured silicon solar cells, all with an SiN_x antireflection coating, under AM 1.5 G illumination. Inset: SEM image of the cell. (B) A photograph of single-crystalline silicon wafers featuring a planar surface (left), reconstructed BSi (middle), and pyramidal texture (right). (C) Reflectance and (D) transmittance of the planar, pristine BSi, and reconstructed BSi samples passivated with an SiO_X/poly-Si (n) stack. The inset in (D) shows schematic illustrations of light trapping in planar and reconstructed BSi samples. ((A) Reproduced with permission from [43]. Copyright 2012 Springer Nature. (B), (C), and (D) Adapted with permission from [44]. Copyright 2022 Elsevier.)

Recent work [44] has demonstrated the first monolithic perovskite/silicon tandem solar cells featuring BSi subcells with TOPCon structures (figure 7.6(B)). Surface reconstruction and alkali treatment enhance passivation and reduce surface recombination while preserving the light-trapping capabilities of BSi. These advancements have led to a certified 28.2% efficiency, one of the highest reported for monolithic perovskite/silicon tandems with TOPCon or double-side-textured silicon bottom cells.

Reflection spectrum (figure 7.6(C)) measurements show that pristine BSi significantly reduces reflectance compared to planar silicon due to its gradual refractive index transition. After surface reconstruction, the average reflectance of BSi slightly increases to 6% but remains lower than the 14% of pyramidal textures, indicating maintained impedance matching. Additionally, increased transmittance (figure 7.6 (D)) at NIR wavelengths suggests enhanced optical paths and stronger infrared scattering (the inset of figure 7.6(D)) [44].

These results demonstrate that combining TOPCon cells with surface reconstruction maintains the light-trapping properties of BSi while improving passivation, making it highly suitable for tandem solar cell applications.

7.6.2 Black silicon in photodetectors

BSi has been increasingly utilized in photodetectors due to its nanostructured surface, which enhances light absorption and detection efficiency, especially in the infrared spectrum [45].

Techniques such as MACE improve its light-trapping capabilities, while ultrafast laser hyperdoping enhances its optoelectronic performance for silicon photonics [46]. Research also shows that femtosecond laser-induced BSi boosts photodetector efficiency and serves as a substrate for surface-enhanced Raman scattering, highlighting its versatility in optical sensing [47]. These advancements emphasize the potential of BSi in developing high-sensitivity photodetectors across a broad wavelength range. In a related study [48], a sub-bandgap NIR silicon photodetector using a BSi/Ag nanoparticle (Ag-NPs) Schottky junction for integrated photonics is demonstrated (figure 7.7(A)). The device combines inner photoemission, light trapping, and plasmon-enhanced absorption to detect NIR light below the Si bandgap. Fabricated via wet etching, it achieves enhanced responsivities of 0.277 $mA\,W^{-1}$ at 1319 nm and 0.226 $mA\,W^{-1}$ at 1550 nm under 3 V bias, outperforming planar-Si/Ag designs. Sub-bandgap light reaches the rear BSi/Ag-NPs junction, where plasmonic excitation drives photocurrent generation.

A recently developed photodetector [49] using sulfur-hyperdoped silicon, fabricated via femtosecond laser processing, demonstrates high responsivity ($>10\ A\,W^{-1}$), exceptional detectivity ($>2 \times 10^{12}$ Jones), and broad detection sensitivity (400–1200 nm) across a wide temperature range (10 K–300 K) (figures 7.7(B), (C)). Hall measurements reveal hyperdoped carriers at sub-bandgap levels, ensuring strong carrier activity. The responsivity peaks at 49 $A\,W^{-1}$ at 150 K, starts at 30.8 $A\,W^{-1}$ at 10 K, and declines to 11.4 $A\,W^{-1}$ at room temperature, indicating

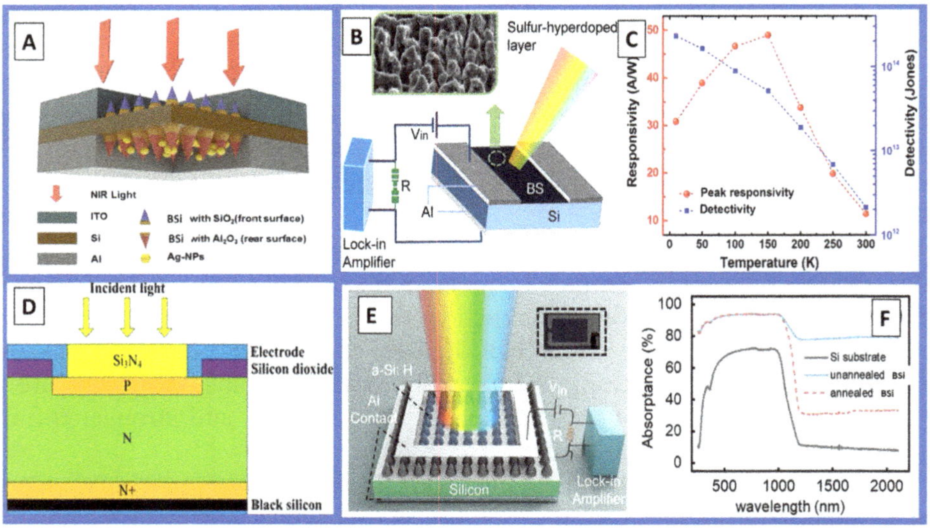

Figure 7.7. (A) A schematic of a BSi/Ag-NP Schottky photodetector. (B) A schematic diagram of the fabricated BSi photodetector, the SEM image reveals the femtosecond laser-modified surface. (C) The temperature dependence of peak responsivity and detectivity in a sulfur-hyperdoped BSi photodetector. (D) The structure of a Si-PIN photodetector with microstructure BSi at the back surface. (E) A cross-sectional schematic diagram of the photodetector, with the inset displaying a photograph of the completed device. (F) The absorptance spectra, compared with untreated silicon, before and after annealing conditions: solid blue line for pre-annealed, dashed red line for post-annealed. ((A) Reproduced from [48]. Copyright 2019 Optica Publishing Group. OSA Open Access. (B) and (C) Reproduced with permission from [49]. Copyright 2021 Elsevier. (D) Reproduced from [50]. Copyright 2018 Springer Nature, CC BY 4.0. (E) and (F) Reproduced with permission from [51]. Copyright 2020 Wiley.)

optimal performance at 150 K (figure 7.7(C)). These properties make it promising for multifunctional optoelectronic applications [49].

Figure 7.7(D) shows a Si-PIN photodetector with microstructured BSi on the back surface, achieving a responsivity of 0.53 A W^{-1} at 1060 nm. Zhong *et al* [50] utilized deep reactive ion etching (DRIE) and plasma immersion ion implantation (PIII) to fabricate this structure on c-Si. Sulfur doping narrows the bandgap and enhances light absorption, especially in the NIR range (800–2000 nm), with a peak absorptance of 83%. This advancement highlights the potential of microstructured BSi for NIR photodetection applications [50].

Figure 7.7(E) presents a schematic of a BSi photodetector, with the inset showing the top view of the device. The sample appears black with minimal reflection, and an Al contact surrounds it [51]. The photodetector exhibits a broad spectral response from 400 to 1600 nm, achieving a peak responsivity of 1097.60 A W^{-1} at 1080 nm under 20 V, the highest reported for silicon photodetectors. This highlights the potential of BSi in infrared detection, night vision imaging, and fiber-optic communication.

Figure 7.7(F) demonstrates that BSi exhibits high photon absorption properties across the visible to IR range. Unannealed BSi demonstrates ~80% absorptance in the 1100–2100 nm range, which remains stable under post-annealing conditions. This high absorptance is primarily attributed to surface microstructures created by femtosecond laser ablation and the intermediate energy levels induced by hyper-doped impurities. However, annealing reduces IR absorption due to thermal diffusion of these impurities. The broad-spectrum high-absorption properties of BSi are key to its exceptional photoresponse [51].

7.6.3 Black silicon in biosensors

BSi has garnered significant attention in the field of biosensors, owing to its unique optical properties and nanostructured surface, which enhance its functionality in various sensing applications.

7.6.3.1 Biosensors based on surface-enhanced Raman spectroscopy

BSi has been extensively explored over the past thirty years as a substrate for surface-enhanced Raman spectroscopy (SERS)-based biosensors, with successful proof-of-concept experiments. The addition of noble metal nanoparticles, such as gold or silver, further boosts sensitivity, enabling the detection of biomolecules, pathogens, and chemicals. These features make BSi-based SERS biosensors ideal for medical diagnostics, environmental monitoring, and food safety applications. Ongoing research aims to optimize the BSi surface morphology to further improve SERS efficiency, making it a promising platform for biosensing applications [52].

The review in [52] highlights the recent progress in fabricating BSi SERS-based biosensors, focusing on advancements in the design of surface morphology and analysing the relationship between microstructuring and the efficiency and sensitivity of SERS. BSi is a promising material for SERS biosensors due to its several key advantages, such as the following: (a) enhanced SERS sensitivity, (b) providing a 3D matrix for uniform plasmonic structure distribution, (c) shifting localized surface plasmon resonance (LSPR) to the NIR, (d) stability and cost-efficiency, (e) easily produced with controllable surface geometry, (f) it can be hydrophobic for analyte concentration or hydrophilic for biocompatibility, and (g) easy integration into existing devices. These attributes make BSi a strong candidate for practical SERS sensing applications.

The study in [53] introduces a rapid, cost-effective method for fabricating Au-coated BSi (BSi/Au) SERS-active substrates with a 10^6 enhancement factor. Using RIE and gold sputtering, the process creates a lace-like silicon surface with uniform gold islands (inset (a) in figure 7.8). The substrates exhibit high uniformity (<6% signal variation), long-term stability (3% signal reduction in one month, 40% in 20 months), and reusability after oxygen plasma cleaning. A case study on doxorubicin (DOX; a widely used anticancer drug) detection (10^{-9}–10^{-4} M), showed a highly reproducible SERS spectrum [53]. These durable, low-cost substrates hold strong potential for scientific and healthcare applications.

The SERS spectra of DOX recorded at three points showed minimal intensity fluctuations.

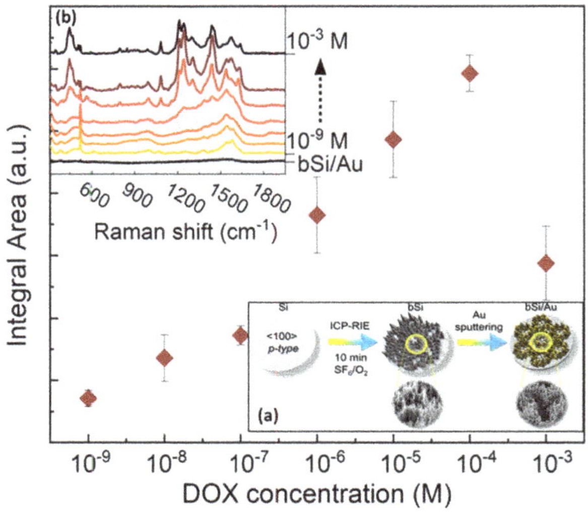

Figure 7.8. Calibration curve for determining DOX concentration using the BSi/Au substrate. Inset (a): Steps of BSi/Au SERS substrate synthesis. Inset (b): SERS spectra of DOX at varying concentrations (1.0 nM to 1.0 mM). (Adapted from [53]. Copyright 2023 American Chemical Society, CC BY 4.0.)

The 1150–1700 cm^{-1} range, corresponding to ring stretch [54], was used to generate a concentration calibration curve (figure 7.8). At DOX concentrations above 10^{-3} M, deviations from monotonicity occurred due to DOX molecular clustering at high concentrations, causing spectral alterations and reduced SERS signals from analyte layers distancing from plasmonic structures. Thus, the effective DOX detection range using BSi is $10^{-9}$and 10^{-4} M [53].

These developments highlight the potential of BSi in advancing optical biosensors, offering enhanced performance through its nanostructured surface and tailored optical characteristics. This offers opportunities to refine and develop new designs for affordable, scalable, and multifunctional sensors.

7.7 Summary

The chapter provides a comprehensive overview of the unique optical properties of black silicon (BSi) and the advanced simulation methods used to analyse them. By enhancing light trapping and reducing reflectance through its distinct surface morphology, BSi demonstrates significant potential for improving the performance of optoelectronic devices. The brief exploration of photoluminescence adds to the understanding of its optical behavior. Furthermore, the diverse applications of BSi in photovoltaics, photodetectors, and biosensors, particularly in enhancing sensitivity for surface-enhanced Raman spectroscopy based biosensors, emphasize its versatility and importance in emerging technologies.

References

[1] Fan Z *et al* 2021 Recent progress of black silicon: from fabrications to applications *Nanomaterials* **11** 41

[2] Liu X *et al* 2014 Black silicon: fabrication methods, properties and solar energy applications *Energy Environ. Sci.* **7** 3223–63

[3] Nguyen K *et al* 2011 Black silicon with sub-percent reflectivity: influence of the 3D texturization geometry *16th Int. Solid-State Sensors, Actuators and Microsystems Conf.* (Piscataway, NJ: IEEE) pp 354–7

[4] Spinelli P, Verschuuren M and Polman A 2012 Broadband omnidirectional antireflection coating based on subwavelength surface Mie resonators *Nat. Commun.* **3** 692

[5] Zhao Z *et al* 2023 Black silicon for near-infrared and ultraviolet photodetection: a review *APL Mater.* **11** 021107

[6] Iandolo B *et al* 2019 Black silicon with ultra-low surface recombination velocity fabricated by inductively coupled power plasma *Phys. Status Solidi—Rap. Res. Lett.* **13** 1800477

[7] Uddin S, Hashim M R and Pakhuruddin M Z 2022 Broadband light absorption enhancement in nanoporous black silicon synthesized by aluminium-catalyzed chemical etching *Opt. Mater.* **134** 113111

[8] Marthi S R, Sekhri S and Ravindra N M 2015 Optical properties of black silicon: an analysis *JOM* **67** 2154–9

[9] Palik E D 1998 *Handbook of Optical Constants of Solids* **vol 3** (New York: Academic)

[10] Fox M 2010 *Optical Properties of Solids* **vol 3** (Oxford: Oxford University Press)

[11] Johnson P B and Christy R W 1972 Optical constants of the noble metals *Phys. Rev.* B **6** 4370–9

[12] Born M, Wolf E and Bhatia A B 1999 *Principles of Optics: Electromagnetic Theory of Propagation, Interference and Diffraction of Light* (Cambridge: Cambridge University Press)

[13] Hecht E 2017 *Optics* (Pearson Education)

[14] Sze S M, Li Y and Ng K K 2021 *Physics of Semiconductor Devices* (New York: Wiley)

[15] Bohren C F and Huffman D R 2008 *Absorption and Scattering of Light by Small Particles* (New York: Wiley)

[16] Zhang Z *et al* 2019 Black silicon with order-disordered structures for enhanced light trapping and photothermic conversion *Nano Energy* **65** 103992

[17] Zhang Z *et al* 2020 Ultralow broadband reflectivity in black silicon via synergy between hierarchical texture and specific-size Au nanoparticles *Adv. Opt. Mater.* **8** 2000668

[18] Hirsch J *et al* 2016 Optoelectronic properties of black-silicon generated through inductively coupled plasma (ICP) processing for crystalline silicon solar cells *Appl. Surf. Sci.* **374** 252–6

[19] Xia Y *et al* 2012 X-ray photoelectron spectroscopic studies of black silicon for solar cell *J. Electron. Spectrosc. Relat. Phenom.* **184** 589–92

[20] Wang H *et al* 2013 Anisotropic optical properties of silicon nanowire arrays based on the effective medium approximation *Int. J. Therm. Sci.* **65** 62–9

[21] Khardani M, Bouaïcha M and Bessaïs B 2007 Bruggeman effective medium approach for modelling optical properties of porous silicon: comparison with experiment *Phys. Status Solidi* C **4** 1986–90

[22] Elsayed A A *et al* 2018 Optical modeling of black silicon using an effective medium/multilayer approach *Opt. Express* **26** 13443–60

[23] Stroud D 1998 The effective medium approximations: some recent developments *Superlattices Microstruct.* **23** 567–73

[24] Zhang D, Cherkaev E and Lamoureux M P 2011 Stieltjes representation of the 3D Bruggeman effective medium and Padé approximation *Appl. Math. Comput.* **217** 7092–107

[25] Branz H M *et al* 2009 Nanostructured black silicon and the optical reflectance of graded-density surfaces *Appl. Phys. Lett.* **94** 231121

[26] Bett A J *et al* 2016 Wave optical simulation of the light trapping properties of black silicon surface textures *Opt. Express* **24** A434–45

[27] Shuba M V *et al* 2015 Adequacy of the rigorous coupled-wave approach for thin-film silicon solar cells with periodically corrugated metallic backreflectors: spectral analysis *J. Opt. Soc. Am.* A **32** 1222–30

[28] Moharam M G *et al* 1995 Stable implementation of the rigorous coupled-wave analysis for surface-relief gratings: enhanced transmittance matrix approach *J. Opt. Soc. Am.* A **12** 1077–86

[29] Lalanne P and Jurek M P 1998 Computation of the near-field pattern with the coupled-wave method for transverse magnetic polarization *J. Mod. Opt.* **45** 1357–74

[30] Tucher N, Gebrewold H T and Bläsi B 2018 Field stitching approach for the wave optical modeling of black silicon structures *Opt. Express* **26** A937–45

[31] Ma S *et al* 2018 A theoretical study on the optical properties of black silicon *AIP Adv.* **8** 035010

[32] Yablonovitch E 1982 Statistical ray optics *J. Opt. Soc. Am.* **72** 899–907

[33] Ayvazyan G, Ayvazyan K and Aghabekyan A 2021 FDTD modeling of antireflection black silicon layers for solar cells applications *Proc. of the 13th Int. Conf. on Comp. Sci. and Inf. Techn*

[34] Sullivan D M 2013 *Electromagnetic Simulation Using the FDTD Method* (New York: Wiley)

[35] Matthias K *et al* 2010 Optical modeling of needle like silicon surfaces produced by an ICP-RIE process *Proc. SPIE* **7725** 772505

[36] Serpengüzel A *et al* 2008 Luminescence of black silicon *J. Nanophotonics* **2** 021770

[37] Lv J *et al* 2018 Review application of nanostructured black silicon *Nanoscale Res. Lett.* **13** 110

[38] Chen T *et al* 2011 Luminescence of black silicon fabricated by high-repetition rate femtosecond laser pulses *J. Appl. Phys.* **110** 073106

[39] Ragavendran V *et al* 2019 Influence of metal assisted chemical etching time period on mesoporous structure in as-cut upgraded metallurgical grade silicon for solar cell application *J. Mater. Sci., Mater. Electron.* **30** 8676–85

[40] Lü Q *et al* 2013 Strong infrared photoluminescence from black silicon made with femtosecond laser irradiation *Opt. Lett.* **38** 1274–6

[41] Bellini E 2022 New kind of black silicon shows improved light-trapping properties *PV Magazine—Photovoltaics Markets and Technology* https://www.pv-magazine.com/2022/07/25/new-kind-of-black-silicon-shows-improved-light-trapping-properties/

[42] Gasparyan F 2023 Optical properties of the crystalline silicon-black silicon-perovskite tandem solar cells *Adv. Mater. Sci. Technol.* **5** 0517626

[43] Oh J, Yuan H-C and Branz H M 2012 An 18.2%-efficient black-silicon solar cell achieved through control of carrier recombination in nanostructures *Nat. Nanotechnol.* **7** 743–8

[44] Ying Z *et al* 2022 Monolithic perovskite/black-silicon tandems based on tunnel oxide passivated contacts *Joule* **6** 2644–61

[45] Alsolami A *et al* 2024 Recent advances in black silicon surface modification for enhanced light trapping in photodetectors *Appl. Sci.* **14** 9841

[46] Huang S *et al* 2024 Ultrafast laser hyperdoped black silicon and its application in photodetectors: a review *Phys. Status Solidi* A **221** 2400127

[47] Wang X *et al* 2024 Femtosecond laser-induced black silicon: a dual application in photodetection and surface-enhanced Raman scattering *Opt. Express* **32** 36362–80
[48] Hu F *et al* 2019 Black silicon Schottky photodetector in sub-bandgap near-infrared regime *Opt. Express* **27** 3161–8
[49] Jin X *et al* 2021 High-performance black silicon photodetectors operating over a wide temperature range *Opt. Mater.* **113** 110874
[50] Zhong H *et al* 2018 Enhanced near-infrared absorber: two-step fabricated structured black silicon and its device application *Nanoscale Res. Lett.* **13** 316
[51] Huang S *et al* 2020 Black silicon photodetector with excellent comprehensive properties by rapid thermal annealing and hydrogenated surface passivation *Adv. Opt. Mater.* **8** 1901808
[52] Padrez Y and Golubewa L 2024 Black silicon surface-enhanced raman spectroscopy biosensors: current advances and prospects *Biosensors* **14** 453
[53] Golubewa L *et al* 2023 Stable and reusable lace-like black silicon nanostructures coated with nanometer-thick gold films for SERS-based sensing *ACS Appl. Nano Mater.* **6** 4770–81
[54] Gautier J *et al* 2013 SERS spectroscopic approach to study doxorubicin complexes with Fe^{2+} ions and drug release from SPION-based nanocarriers *Analyst* **138** 7354–61

IOP Publishing

Black Silicon
Technology, opportunities and challenges
N M Ravindra and Priyanka Singh

Chapter 8

Electrical properties

This chapter explores the electrical properties of black silicon (BSi), emphasizing recombination mechanisms, carrier concentration, mobility, and electrical conductivity. It examines various characterization techniques such as current–voltage (I–V) characteristics, spectral response (SR), minority carrier lifetime (τ), and quantum efficiency (QE). Additionally, various passivation schemes aimed at minimizing the recombination and improving the device performance are discussed. The chapter also highlights applications in optoelectronics, including photovoltaic devices, sensors, and detectors.

8.1 Introduction

Black silicon (BSi) is a nanostructured form of silicon characterized by its unique surface morphology, which results in superior optical and electronic properties compared to conventional silicon (Si). Initially developed for photovoltaic (PV) applications, due to its excellent light absorption capabilities, BSi has garnered attention for its distinctive electrical and optical properties, making it a promising material for various electronic and optoelectronic devices [1].

BSi is commonly produced using techniques such as chemical etching, reactive ion etching (RIE), metal-assisted chemical etching (MACE), electrochemical etching (ECE), plasma immersion ion implantation (PIII), or femtosecond laser irradiation [2], as discussed in chapter 4. The micro- and nanostructures, formed through these techniques, enhance its surface area and exhibit distinct electronic properties [3] (addressed in chapter 6), with atomic interactions and microstructure affecting electrical conductivity and thermal properties (chapter 5) [3]. The morphology, including the size, shape, and distribution of these nanostructures, plays a crucial role in determining the electrical properties of BSi. While BSi offers excellent light absorption, its surface nanostructures cause defects, increased surface recombination, and poor electrical performance. Challenges to the use of BSi include the following: (i) more dangling bonds reducing carrier lifetimes, (ii) enhanced Auger

doi:10.1088/978-0-7503-3247-7ch8

recombination from high dopant concentrations, and (iii) weak metal contact adhesion. These factors lead to lower efficiencies in BSi-based solar cells compared to advanced technologies [4–6]. Liu *et al* [1] and Toor *et al* [7] have reviewed these challenges and proposed solutions. Additionally, residual stress and interface quality play a crucial role in electron transport, affecting properties such as carrier lifetimes, recombination velocities, and band structure. Some BSi geometries have achieved carrier lifetimes in the millisecond range [4]. BSi exhibits strong near-band edge absorption due to light scattering at surface structures, leading to total internal reflection. Deep BSi structures, with steep sidewalls and optimal inter-needle spacing, enhance light absorption through multiple internal reflections, while shallow and intermediate structures show similar optical performance despite differences in electronic properties. With minimized plasma damage and effective surface passivation, BSi can achieve carrier lifetimes that are comparable to polished wafers.

8.2 Recombination mechanisms

The nanostructures in BSi significantly influence the recombination dynamics [8]. As illustrated in figure 8.1, primarily four recombination mechanisms exist: (a) radiative recombination, (b) Shockley–Read–Hall (SRH; bulk defect) recombination, (c) surface recombination, and (d) Auger recombination [9].

8.2.1 Radiative recombination

Radiative recombination, in which electrons and holes recombine to emit photons (figure 8.1(a)), is the inverse of optical absorption. In indirect bandgap semiconductors such as silicon, it requires phonon participation to conserve momentum, making it negligible compared to other recombination processes [8, 9]. This effect is even less significant in BSi due to the dominance of non-radiative processes.

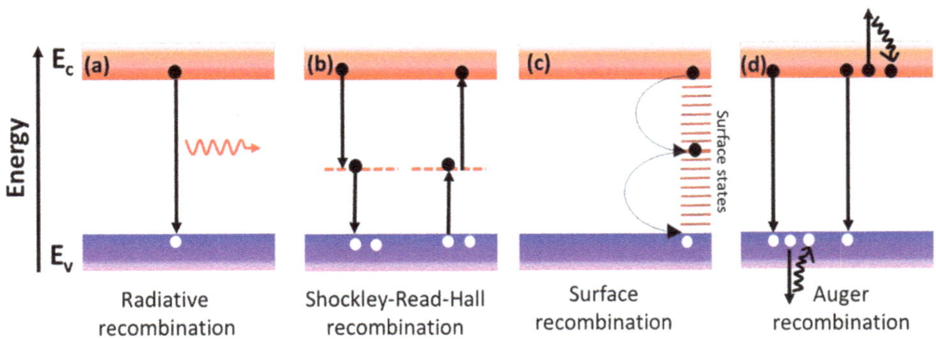

Figure 8.1. Schematic illustration of fundamental semiconductor recombination mechanisms, in which electrons (black circles) and holes (white circles) move in a direction opposite to each other. (a) Radiative recombination is the reverse of photon absorption. (b) SRH recombination occurs via defect-induced energy levels. (c) Surface recombination occurs through surface states. (d) Auger recombination excites an additional carrier by absorbing the released energy.

8.2.2 Shockley–Read–Hall recombination

SRH recombination, or trap-assisted recombination, occurs via defect-induced energy levels within the bandgap, and is caused by impurities such as iron (Fe) or copper (Cu) or crystallographic imperfections. This two-step process involves an electron relaxing from the conduction band to a defect level, then to the valence band to recombine with a hole (figure 8.1(b)) [8, 9]. SRH recombination is significant in silicon solar cells due to defects from crystal growth or contamination, but can be mitigated through gettering or trap passivation, such as hydrogen release from silicon nitride coatings. In BSi, nanostructuring increases defect sites, enhancing SRH recombination, which can be reduced by high-temperature annealing and passivation.

8.2.3 Surface recombination

Defects and impurities on the surface of the semiconductor enhance surface recombination (figure 8.1(c)) [8, 9]. The surface of BSi, in which the crystal lattice is significantly disrupted, experiences especially high recombination rates. Consequently, it is a critical issue in BSi due to its high surface state density from nanostructures, leading to increased surface defects, dangling bonds, and carrier trapping, which reduce carrier lifetime and device efficiency. High aspect ratio nanostructures further exacerbate recombination by introducing more defects and bandgap alterations [10, 11]. Passivation methods, such as the use of thin dielectric layers, are used to mitigate recombination losses and improve electrical performance. In order to mitigate surface recombination, two strategies are employed: (i) reducing interface trap density with passivation layers, and (ii) minimizing minority carrier concentration at the surface through doping (e.g. back surface field) or fixed-charge dielectric layers (e.g. aluminum oxide on p-doped surfaces) that create electric fields to repel carriers.

Surface recombination in BSi can be reduced through surface passivation, layer thickness optimization, and surface cleaning. Passivation decreases dangling bonds and surface defects, enhancing carrier lifetime and solar cell efficiency. Liu *et al* [12] showed that passivation significantly improves the quantum efficiency of BSi cells to levels that are comparable to conventional cells. Similarly, Repo *et al* [13] found that aluminum oxide (Al_2O_3) passivation increases carrier lifetimes to levels matching those of conventional multicrystalline silicon solar cells.

Optimizing the thickness of the BSi layer is key to reducing the surface recombination. While taller nanostructures improve light absorption, they also increase the surface area and recombination. Otto *et al* [4] found that BSi etched to 600 nm, with a 100 nm Al_2O_3 passivation layer, achieved longer carrier lifetimes, comparable to conventional cells, outperforming shallower (500 nm) and deeper (1400 nm) etching. This highlights the need for an optimal BSi thickness that balances light absorption and minimizes recombination [14].

Surface cleaning effectively reduces surface defects and surface recombination in BSi solar cells. Xiao *et al* [15] showed that BSi cells, fabricated using PIII, performed poorly when compared with acid-textured cells, but defect removal etching improved their performance beyond that of the acid-textured cells. Similarly, Su *et al* [16] found that defect removal in BSi cells, made with MACE, reduced recombination sites. Combining defect removal with passivation can significantly minimize surface recombination in BSi solar cells.

8.2.4 Auger recombination

Auger recombination is a non-radiative process in which recombination energy is transferred to a third carrier (electron or hole) instead of being emitted as light [8, 9] (figure 8.1(d)). It becomes significant in highly doped BSi, particularly under high-injection conditions, but can be mitigated by optimizing the doping levels to reduce excessive carrier concentrations. BSi nanostructures intensify both surface and Auger recombination, as the energy from electron–hole recombination is transferred to another conduction band electron, which dissipates the excess energy as phonons [9, 10].

The Auger recombination rate rises with increasing dopant concentrations and follows a well-defined relationship [10]:

$$R_{\text{Auger}} = C_{\text{n}} n^2 p + C_{\text{p}} p^2 n, \tag{8.1}$$

where R_{Auger} represents the Auger recombination rate, C_{n} and C_{p} are constants (for silicon, $C_{\text{n}} \approx C_{\text{p}} \approx 1 \times 10^{-30}\ \text{cm}^6\ \text{s}^{-1}$) that define the material-specific characteristics. The variables n and p represent the concentrations of electrons and holes, respectively. Since these concentrations are usually similar, the recombination rate is proportional to the cube of either of the charge carriers. Studies show that Auger recombination is dominant in highly doped regions, especially in BSi solar cells [17]. In order to minimize this, dopants should be evenly distributed across the wafer. However, the large surface area of BSi results in heavy front-surface doping, leading to high carrier concentrations and a significant increase in the Auger recombination rate near the surface. Auger recombination in BSi solar cells can be reduced by controlling the dopant profile using methods such as: (i) lowering the doping temperature, (ii) etching away the top layer with high dopant concentration, and (iii) doping the substrate before BSi etching. Zhong *et al* [18] have reported studies wherein they lowered the doping temperature, reducing dopant concentration due to a lower diffusion coefficient, which also improved the device quantum efficiency. Oh *et al* [17] used tetramethylammonium hydroxide (TMAH) etching to remove the highly doped top BSi layer, enhancing the carrier lifetime and internal quantum efficiency despite increased surface reflectance. Shen *et al* [19] and Wang *et al* [20] showed that doping before BSi etching reduces Auger recombination and improves conversion efficiency compared to cells etched before doping.

In addition, a study of recombination mechanism in ion-implanted and diffused BSi emitters revealed that diffused emitters experience higher Auger recombination due to increased doping, whereas ion implantation allows precise doping control regardless of the surface area [8]. Figure 8.2 presents electrochemical capacitance–voltage (ECV) measured doping profiles of ion-implanted and diffused BSi emitters. In ion-implanted emitters, surface doping decreases [21, 22] from $(4.5 \pm 0.1) \times 10^{19}\ \text{cm}^{-3}$

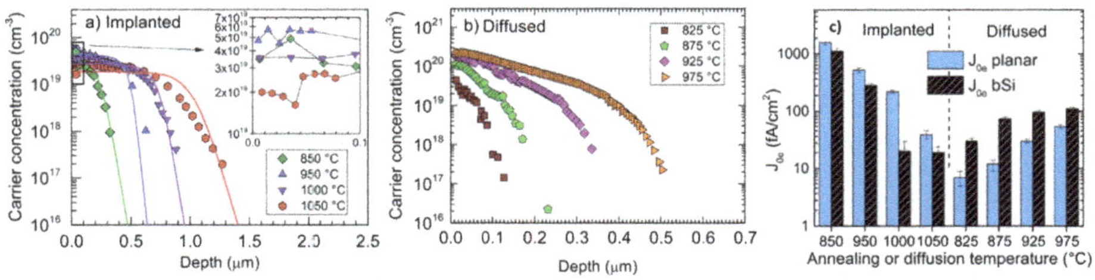

Figure 8.2. The doping profiles of (a) implanted and (b) diffused planar emitters, measured using the ECV technique. Simulated profiles from the ICECREM software (solid lines) align with experimental data. The inset in (a) shows surface doping values across anneal temperatures, with solid lines as visual guides. (c) A comparison of emitter saturation current density (J_{0e}) values in planar and BSi emitters processed by implantation and diffusion. (Reproduced with permission from [8]. Copyright 2017 AIP Publishing.)

to $(1.90 \pm 0.07) \times 10^{19}$ cm^{-3} as anneal temperature rises, while emitter depth increases from 0.7 to 1.5 μm (figure 8.2(a)). In contrast, surface doping in diffused emitters rises from 4.5×10^{19} to 2.2×10^{20} cm^{-3} at higher temperatures (figure 8.2(b)), leading to higher Auger recombination. Furthermore, doping profiles obtained from the ICECREM (Integrated Circuits and Electronics group Computerized Remedial Education and Mastering) [23] process simulator closely match those from experimental ECV measurements (figure 8.2(a)). Additionally, implanted emitters in BSi achieve lower emitter saturation current density (J_{0e}) than planar emitters (figure 8.2 (c)), while diffused BSi emitters show higher J_{0e} due to excessive doping [17]. The increased J_{0e} in diffused BSi emitters is linked to higher Auger recombination caused by elevated doping levels. It has been observed that higher annealing temperatures reduce surface recombination, achieving low J_{0e} (20 ± 5 fA cm^{-2} at 170 Ω sq^{-1}). Ion implantation enables precise doping control while maintaining BSi morphology and low reflectance. Bulk recombination decreases at lower anneal temperatures, and surface recombination remains low. Beyond 1000 °C, J_{0e} is no longer surface-limited, highlighting the potential of ion implantation for fabricating optimized BSi emitters [8].

8.3 Charge carrier dynamics

BSi exhibits unique charge carrier dynamics due to its nanostructured surface, which significantly influences its optoelectronic properties. The high surface area and deep subwavelength structures enhance light absorption but also introduce additional surface states that impact carrier recombination and transport. Understanding and optimizing charge carrier dynamics in BSi is essential for advancing its applications in the next-generation energy and optoelectronic devices.

8.3.1 Carrier concentration and mobility

The nanostructured surface of BSi affects charge carrier concentration and mobility (μ) due to increased surface area and surface states. Optimized fabrication can reduce defects and enhance mobility, making BSi suitable for high-performance optoelectronics applications. Carrier concentration and mobility are important parameters that define the electrical properties of BSi. Carrier concentration depends on doping level (n-type or p-type), surface defects that create trap states, and passivation to mitigate these defects [24, 25]. Typical carrier concentrations range from 10^{15} to 10^{18} cm^{-3}, depending on doping and fabrication methods.

Mobility in BSi is lower than in bulk crystalline silicon due to factors such as surface scattering, defects acting as recombination centers, and quantum confinement at extremely small scales. While bulk silicon has electron and hole mobilities of $\sim$1350 and $\sim$450 $cm^2\ V^{-1}\ s^{-1}$, respectively, BSi typically exhibits reduced electron mobility in the range of 100–500 $cm^2\ V^{-1}\ s^{-1}$, depending on the fabrication quality and passivation.

8.3.2 Electrical conductivity

The electrical conductivity of BSi is influenced by doping, temperature, and surface morphology. Intrinsic electrical conductivity arises from thermally generated carriers, while extrinsic electrical conductivity depends on doping levels. Nanostructures modify the band structure, affecting the electrical conductivity of BSi. Doping with group V (n-type) or group III (p-type) elements enhances charge carrier concentration, but surface defects can reduce mobility. Furthermore, BSi may exhibit anisotropic electrical conductivity due to its nanostructured surface, causing directional variations in its electrical properties. This feature is advantageous for developing anisotropic electronic devices and sensors. Additionally, the high surface-to-volume ratio of BSi amplifies the impact of surface states and defects on electrical behavior. These states can trap carriers (figure 8.1(c)), increasing recombination and affecting the electrical conductivity of BSi.

The electrical properties of Au-doped BSi, fabricated via femtosecond laser irradiation, were studied in [24] using Hall effect measurements with the van der Pauw technique [26]. Figure 8.3 presents the carrier mobility and sheet electron density (Ns) of annealed Au-doped BSi samples [24].

Au doping in n-type Si reduced Ns due to self-compensation between deep donor and acceptor levels. The conduction type depended on the substrate, with the Au-doped layer in n-type Si retaining n-type conductivity. The carrier electron density remained close to that of the substrate (3.26×10^{10} cm^{-2}), ranging from 2.66×10^{10} to 5.73×10^{10} cm^{-2}, due to Au impurities acting as both donors and acceptors [27]. In p-type Si (1500 Ω·cm, 350 μm figure 8.3(b)), Au doping resulted in p-type conductivity with a higher hole carrier density (6.23×10^{11} to 1.21×10^{12} cm^{-2}) than the substrate (5.42×10^{10} cm^{-2}), and it can be attributed to increased acceptor density in the Au-doped layer [24]. These measurements confirmed that both p-type and n-type Au-doped Si can be achieved by selecting the appropriate substrate type.

At higher laser fluence, increased damage to the Au film reduces Au dopant concentration, lowering carrier density in Au-doped BSi. This decreases impurity scattering, slightly increasing carrier mobility, as shown in figure 8.3. The variation in

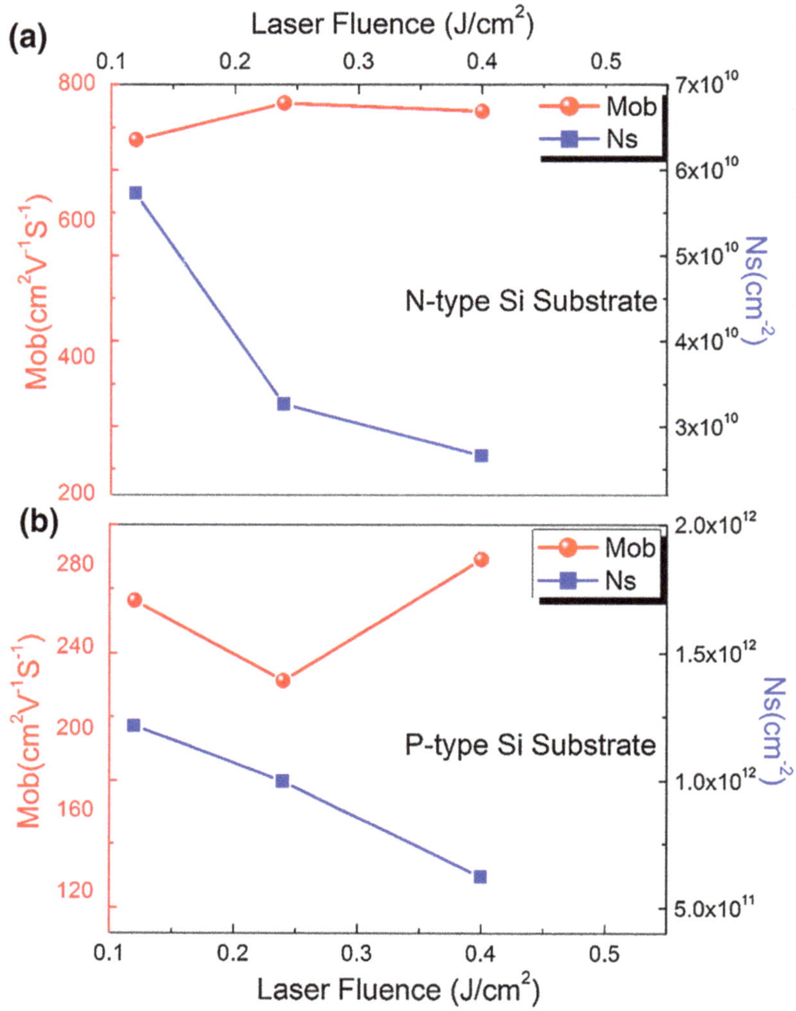

Figure 8.3. (a) Sheet carrier density (Ns) and mobility (Mob) of Au-doped BSi fabricated at varying laser fluences using an N-type Si substrate. (b) Ns and Mob of Au-doped BSi fabricated at varying laser fluences using a P-type Si substrate. (Reproduced with permission from [24]. Copyright 2017 Springer Nature.)

mobility (figure 8.3) results from surface damage and impurity scattering, with electron mobility (719–773 $cm^2\ V^{-1}\ s^{-1}$) exceeding the hole mobility (224–283 $cm^2\ V^{-1}\ s^{-1}$) [24].

Despite its lower mobility, excellent light absorption in BSi enhances photocurrent generation in PV devices. However, increased surface recombination can reduce efficiency unless properly passivated. Its strong light-trapping capabilities make it valuable for low-light detection despite mobility limitations. Optimizing doping methods and implementing passivation techniques enhances carrier lifetime and overall efficiency. Effective surface passivation and precise fabrication control are essential for minimizing losses and maximizing the device performance.

8.4 Characterization techniques

This section covers important characterization techniques and equations for evaluating the electrical properties of BSi devices, emphasizing performance analysis of BSi solar cells and photodetectors. Characterization primarily involves current–voltage (I–V), spectral response (SR), and quantum efficiency (QE) measurements including lifetime (τ) measurements. The following section presents the essential equations that are used in these methods to assess material quality and enhance the overall device performance.

8.4.1 Illuminated current–voltage characteristics

Figure 8.4(a) (inset) illustrates the equivalent circuit of an ideal solar cell [27]. An ideal solar cell is a theoretical PV device with no losses, characterized by the absence of series resistance (R_s), shunt resistance (R_{sh}), and an ideality factor (n) of 1. This section considers the ideal case to provide a fundamental understanding of the PV cell operation. The I–V characteristics of an ideal p–n junction solar cell, under steady-state illumination, can be most simply represented using a single exponential model as

$$I = -I_{ph} + I_o\left\{\exp\left(\frac{qV}{kT}\right) - 1\right\}, \tag{8.2}$$

where I_{ph} represents the photogenerated current, V is the terminal voltage across the junction, k is the Boltzmann constant, and I_o is the saturation current.

8.4.2 Dark current–voltage characteristics

The dark current–voltage (I_D–V) characteristics of an ideal PV cell can be described by substituting $I_{ph} = 0$ in equation (8.2) and can be formulated as

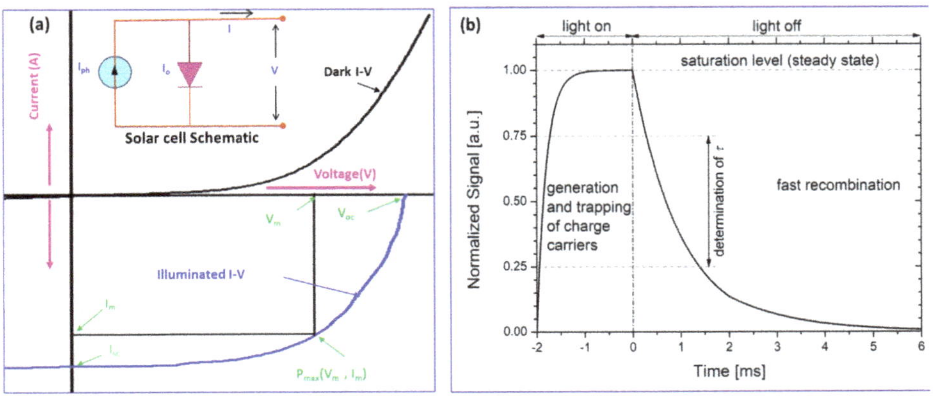

Figure 8.4. (a) Schematic illustration of dark and illuminated current–voltage (I–V) characteristics of a solar cell. The inset in (a) shows a schematic of an ideal p–n junction solar cell. (b) Depiction of the minority carrier lifetime measurement. ((b) Reproduced from [28]. Copyright 2019 MDPI, CC BY 4.0.)

$$I = I_o\left\{\exp\left(\frac{qV}{kT}\right) - 1\right\}. \tag{8.3}$$

I–V measurements for photodiodes provide an analysis of the current response under varying voltage conditions to determine carrier lifetime and recombination rates. By examining I–V curves at different illumination intensities, recombination currents and carrier lifetimes can be extracted. This method is particularly relevant for photodetectors and photodiodes, where lifetime impacts the signal-to-noise ratio.

8.4.3 Solar cell performance parameters

Performance parameters, namely short-circuit current (I_{sc}), open-circuit voltage (V_{oc}), fill factor (FF), and efficiency (η) are used to characterize an illuminated solar cell.

The open-circuit voltage, V_{oc}, is defined as the voltage when terminals are open in the solar cell. By setting the total current, I, to zero in equation (8.2), the relation between V_{oc}, I_o, and I_{ph} for single exponential model is defined as

$$V_{oc} = \frac{kT}{q}\ln\left\{\frac{I_{ph}}{I_o} + 1\right\}. \tag{8.4}$$

V_{oc} is related to I_{ph} and I_o. For a high V_{oc}, a low I_o is absolutely necessary.

The fill factor is defined as the ratio of the maximum power (P_{max}) at maximum power point (MPP) to the product V_{oc} and I_{sc} ($\approx I_{ph}$) and its value is always less than unity and can be expressed as

$$\mathrm{FF} = \frac{P_{max}}{V_{oc}I_{sc}}. \tag{8.5}$$

The condition for MPP and expression for P_{max} is given by

$$\frac{\mathrm{d}P}{\mathrm{d}V} = 0 \Rightarrow \frac{\mathrm{d}(IV)}{\mathrm{d}V} = 0 \tag{8.6}$$

$$P_{max} = I_m V_m, \tag{8.7}$$

where I_m and V_m represent the current and voltage at the MPP (figure 8.4(a)). The output power per unit area of the cell can be determined from the illuminated I–V characteristics by evaluating the area of the rectangle formed between the MPP and the two axes (as shown in figure 8.4(a)). The operating point depends on the load resistance.

The efficiency (η) of a solar cell is defined as the ratio of P_{max} per unit area to P_{in} and is represented as

$$\eta = \frac{P_{max}}{P_{in}.\ \mathrm{Area}}, \tag{8.8}$$

where P_{in} is the intensity of the incident radiation.

In terms of the V_{oc}, J_{sc}, and FF, the efficiency of the cell can be represented as

$$\eta = \frac{V_{oc}.\, J_{sc}.\, \mathrm{FF}}{P_{in}}, \tag{8.9}$$

where J_{sc} is the current density (current per unit area). The short-circuit current density of the cell depends on the given solar spectral irradiance and is given by

$$J_{sc} = q \int_{h\nu=E_g}^{\infty} \frac{\mathrm{d}N_{ph}}{\mathrm{d}h\nu}\, \mathrm{d}\left(h\nu\right), \tag{8.10}$$

where N_{ph} is the incident photon flux. In a practical solar cell, the value of J_{sc} may be limited by reflection losses, ohmic losses (R_s and R_{sh}) and recombination losses.

8.4.4 Spectral response

The spectral response (SR) measurement characterizes the response of a solar cell to the incident light wavelengths and helps to determine the spectral range in which it operates efficiently. It provides insights into material properties and device design. SR is defined as the ratio of the J_{sc} produced under monochromatic illumination at a specific wavelength (λ) to P_{in} at that wavelength.

SR can be expressed mathematically as

$$\mathrm{SR}(\lambda) = \left(\frac{J_{sc}}{P_{in}}\right)_{\lambda}. \tag{8.11}$$

8.4.5 Quantum efficiency

The QE of a solar cell is defined as the ratio of electrons generated in the external circuit to the number of incident photons at a particular wavelength. It is categorized into two types: internal quantum efficiency (IQE) and external quantum efficiency (EQE). EQE is determined by considering all incident photons, whereas IQE accounts only for photons that are not reflected. The experimentally obtained results of EQE help assess how efficiently the solar cells convert the incident photons into electrical current across different wavelengths. This technique provides insights into the SR of the solar cell and helps identify factors affecting light absorption and carrier generation. Conceptually, the SR and QE are closely related.

Mathematically, EQE and IQE can be defined as follows:

$$\mathrm{EQE} = \frac{hc}{q\lambda}\mathrm{SR}(\lambda) \tag{8.12}$$

$$\mathrm{IQE} = \frac{\mathrm{EQE}}{(1 - R_{\lambda})}, \tag{8.13}$$

where h is Planck's constant, c is the velocity of light, and R_{λ} is the reflection coefficient.

8.4.6 Minority carrier lifetime

Minority carrier lifetime (τ) in BSi is crucial for optoelectronic applications, affecting recombination rates and device efficiency [27]. It represents the average time a minority carrier exists before recombining with a majority carrier. Measuring the minority carrier lifetime in BSi is challenging due to increased surface recombination, optical absorption variations, and interference from surface roughness. Various techniques, including transient photoconductance (TPC), quasi-steady-state photoconductance (QSSPC), microwave photoconductance decay (μ-PCD) [29–31], and time-resolved photoluminescence (TRPL) [32], are used to measure carrier lifetime, aiding in material quality assessment and performance optimization.

TPC measures photoconductance decay after an optical pulse to evaluate surface and bulk recombination, aiding in process monitoring and passivation quality assessment in BSi solar cells. QSSPC measures effective minority carrier lifetime (τ_{eff}) by analysing photoconductance decay under steady-state illumination to evaluate the quality of passivation and material properties in high-performance optoelectronic devices [31]. μ-PCD, a non-contact microwave technique, enables spatially resolved carrier lifetime mapping, making it ideal for characterizing BSi morphologies, assessing passivation effects, and optimizing device fabrication while minimizing contamination risk. As shown in figure 8.4(b) [28], in the μ-PCD technique [29], excess carriers are optically generated until a steady-state photoconductivity is established. After switching off the optical excitation, the excess carriers decay, leading to a decrease in photoconductivity. The carrier lifetime is determined from the exponential decay, using linear regression between 75% and 25% of the signal height, as shown in figure 8.4(b) [33]. Photoconductivity variations are monitored through microwave absorption, with the wafer placed in a resonant microwave cavity. Excitation is typically provided by laser diodes [28].

The TRPL technique assists in the analysis of photoluminescence (PL) decay after pulsed excitation to measure carrier recombination time [32]. It is highly sensitive to both surface and bulk recombination, providing insight into carrier dynamics and passivation quality. TRPL is useful for detecting defects, quantifying lifetime variations, and optimizing passivation layers in optoelectronic devices:

$$I(t) = I_0 \exp\left(-t/\tau\right), \tag{8.14}$$

where $I(t)$ is the PL intensity at time t and I_0 is the initial PL intensity at $t = 0$.

A rapid and precise estimation of the minority carrier effective lifetime (τ_{eff}) is crucial in photoconductance measurements [27, 31, 34]. A commonly used simplified equation is given in [34]:

$$\frac{1}{\tau_{\text{eff}}} = \frac{1}{\tau_{\text{bulk}}} + \frac{2S_{\text{eff}}}{W}, \tag{8.15}$$

where τ_{bulk} represents the lifetime of excess charge carriers within the bulk of the semiconductor, and is determined by the recombination processes occurring within the material (radiative, SRH, Auger) [35]. In equation (8.15), W is the wafer thickness and S_{eff} is the surface recombination velocity which describes the rate at which charge carriers recombine at the surface. When both surfaces are identical, the surface recombination velocity is related to the effective minority carrier lifetime by the above relation [36].

The variations in optical absorption can lead to bulk lifetime underestimation and can also interfere with optical-based measurement techniques, impacting accuracy. The strategies to improve measurement accuracy include utilizing high-quality passivation layers to effectively reduce surface recombination [37]. Hydrogenation treatments, such as plasma treatment, enhance carrier lifetime by mitigating surface defects and dangling bonds. In addition, optimizing doping concentrations and profiles help minimize Auger and bulk recombination, leading to improved charge carrier lifetime.

Accurate minority carrier lifetime measurement is essential for optimizing BSi in optoelectronic applications. Techniques such as QSSPC, μ-PCD, and TRPL provide key insights but must account for passivation and surface effects. Effective measurement methods and improved passivation strategies can significantly enhance the device performance.

8.5 Passivation

Passivation is essential for enhancing BSi performance by reducing surface recombination and improving electrical properties, particularly in PV applications. Atomic layer deposition (ALD) deposited aluminum oxide (Al_2O_3) has achieved surface recombination velocities as low as 8 cm s^{-1} on p-type Si float zone (FZ) substrates. Pre-treating BSi surfaces with chemical solutions before ALD has further improved the quality of passivation [14]. Plasma-enhanced chemical vapor deposition (PECVD) is effective for modulated surface textures (MSTs), [38, 39] while numerical simulations optimize field-effect passivation [40].

In c-Si solar cells, silicon dioxide (SiO_2), silicon nitride (SiN_x), and Al_2O_3 are used for passivation to minimize recombination and protect against environmental factors, while hydrogen passivation is vital for amorphous silicon (a-Si) cells. In BSi solar cells, micro/nanostructures intensify recombination, however, applying SiO_2, SiN_x, Al_2O_3, and a-Si helps counteract these effects, improving efficiency by reducing the surface recombination and managing quantum confinement effects.

8.5.1 Silicon nitride (SiN_x) passivation

SiN_x is commonly used to passivate both n-type and p-type silicon solar cell wafers. It is typically deposited using PECVD at around 400 °C with a mixture of ammonia (NH_3) and silane (SiH_4) gases. By adjusting the flow rates of NH_3 and SiH_4, the silicon-to-nitrogen ratio can be controlled. Adding nitrogen (N_2) to the gas mixture lowers the silicon content, improving the etching rate of SiN_x, reducing UV light absorption, and enhancing the minority carrier lifetime [41]. The passivation effect in SiN_x primarily reduces the dangling bonds at the surface, and as N_2 content increases, it creates a high density of positive charges that help passivate n-Si, thereby reducing the surface minority carrier concentration [42]. SiN_x passivation is effective for n-Si and lightly doped p-type Si, and it also serves as an anti-reflective (AR) layer. It has been used to passivate n^+BSi emitters produced by various processes.

8.5.2 Silicon dioxide (SiO_2) passivation

Thermal oxidation is a passivation technique that reduces Si interfacial defects to a range of 10^9 to 10^{10} eV^{-1} cm^{-2}. It can be used for both n-type and p-type Si. A layer of aluminum (Al) is pre-deposited on an oxidized wafer, and during annealing, hydrogen released from the oxidation of Al passivates the SiO_2/Si interface. The remaining Al is then removed by etching. This process achieves low surface recombination rates of 2 and 12 cm s^{-1} [43].

Thermal oxidation of Si occurs at 800 °C to 1100 °C, requiring low-impurity c-Si wafers. To avoid high temperatures, low-temperature wet oxidation (800 °C) or PECVD deposition of SiO_x can be used, though PECVD layers provide weaker passivation than thermally grown SiO_2. The device performance can be improved by capping with amorphous SiN_x or Al_2O_3 layers.

8.5.3 Aluminum oxide (Al_2O_3) passivation

Aluminium oxide (Al_2O_3) is an excellent surface passivation material, deposited using methods such as ALD, PECVD, and pulsed layer deposition (PLD) [44, 45]. ALD is widely used for BSi passivation due to its superior surface conformality. The ALD process uses trimethylaluminum [TMA–$Al(CH_3)_3$] as the aluminum precursor and H_2O as the oxidative agent, forming an Al_2O_3 [46] atomic layer per cycle in a self-limiting reaction. This method provides precise thickness control and high conformality over large areas [44]. When additional oxidation sources such as O_2 or O_3 plasma are introduced, ALD is classified as thermal ALD or plasma-assisted ALD (PE-ALD) [44, 45]. PE-ALD is 30% faster than thermal ALD, offering better quality at lower temperatures and reduced impurity levels.

In PE-ALD, the interfacial SiO_x layer between Al_2O_3 and Si thickens upon annealing, reducing interfacial defects [47]. The defect density at the Al_2O_3/Si interface is $\sim 10^{11}$ $eV^{-1}cm^{-2}$. Al_2O_3 provides strong field-effect passivation due to its high fixed negative charge density ($\sim 10^{13}$ cm^{-2}) [48, 49] leading to low surface recombination velocities. ALD offers excellent thickness control, step coverage, conformality, low defect density, high uniformity, reproducibility, and low deposition temperature [48]. It is also UV-resistant but has a slow growth rate [50]. ALD-Al_2O_3 passivated BSi solar cells achieve an efficiency of 22.1% [41].

8.6 Applications in optoelectronics

BSi is valuable in optoelectronics due to its superior light absorption, reduced reflectance, and enhanced carrier collection. It improves light trapping and passivation in passivated emitter and rear cell (PERC), tunnel oxide passivated contact (TOPCon), and heterojunction solar cells, boosting efficiency. In infrared (IR) and ultraviolet (UV) photodetectors, its high surface area enhances sensitivity for imaging and sensing. BSi also improves carrier injection and light extraction in

Si-based light emitting diodes (LEDs), while enhancing absorption in laser detectors and optical sensors for light detection and ranging (LiDAR) and spectroscopy. Additionally, its modified electronic properties benefit waste heat recovery and hybrid solar applications. The following section will explore BSi applications in photovoltaics and photodetectors, highlighting their electrical properties with examples.

8.6.1 Photovoltaic devices

Superior light absorption and adjustable electrical properties of BSi boost photovoltaic performance by enhancing carrier collection and minimizing reflectivity, resulting in increased solar cell efficiency. BSi serves as an anti-reflection coating (ARC) in solar cells, replacing traditional ARC layers while maintaining the standard silicon solar cell structure. Its textured surface reduces reflectance, enhancing optical properties and potentially improving efficiency. However, previous attempts to integrate BSi into solar cells have resulted in lower efficiencies due to high surface recombination, which limits electrical performance and lowers power conversion efficiency [36]. Despite these challenges, BSi offers potential for improved efficiency and reduced manufacturing costs in solar cells.

In a recent work [51], high-efficiency bifacial TOPCon solar cells have been developed by integrating BSi with a nanopore pyramid emitter using the RIE technique (figure 8.5). The optimization of the Al_2O_3/SiN_x passivation stack and boron diffusion process has enabled the fabrication of large-area (158.75 × 158.75 mm^2) solar cells with J_{sc} of 40.69 mA cm^{-2} and 23.01% efficiency using industrial processes. Figure 8.5(a) presents schematic illustrations of the main steps in the fabrication of BSi TOPCon solar cells. Figure 8.5(b) illustrates the structural diagram of the BSi TOPCon solar cell. The tunnel oxide and n$^+$ poly layer is deposited using a low-pressure chemical vapor deposition (LPCVD) technique, while the front ARC, consisting of ALD-Al_2O_3 and PECVD-SiN_x, provide excellent

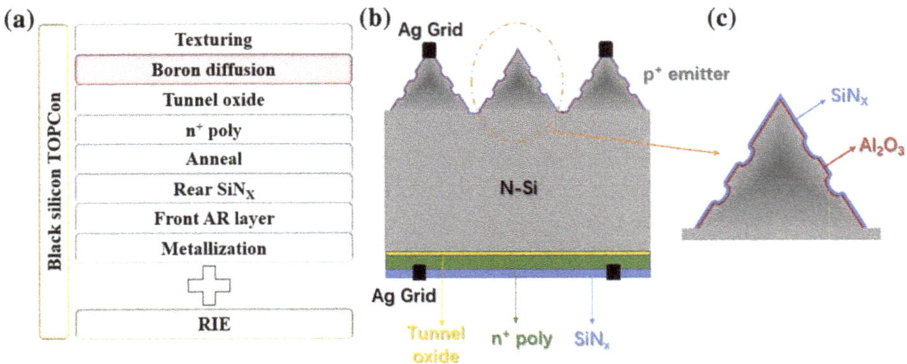

Figure 8.5. Fabrication of BSi TOPCon solar cells. (a) Manufacturing steps of the TOPCon solar cell incorporating an additional RIE process. (b) Structural diagram of the BSi TOPCon solar cell. (c) The nanopore pyramid structure emitter. (Reproduced with permission from [51]. Copyright 2021 Springer Nature.)

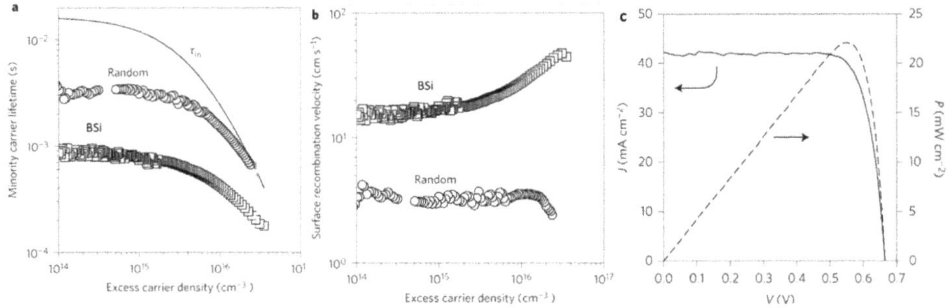

Figure 8.6. (a) Minority carrier lifetime and (b) corresponding surface recombination velocities as functions of excess carrier density. The BSi sample values are represented by squares, while the random pyramid sample values are represented by circles. The solid line in (a) indicates the intrinsic lifetime(τ_{in}). Both samples have a substrate resistivity of 2.6 ± 0.2 Ω cm. (c) The solid line depicts the current density–voltage (*J–V*) characteristics, while the dashed line represents the power–voltage (*P–V*) characteristics of the selected BSi solar cell. (Reproduced with permission from [36]. Copyright 2015 Springer Nature.)

passivation [52]. Figure 8.5(c) depicts the nanopore pyramid structure of emitter which was fabricated using a plasma mixture of O_2, Cl_2, and SF_2 at 13.56 MHz RF plate power of 100 W following silicon texturing [51].

Interdigitated back-contacted (IBC) BSi solar cells have achieved over 22% efficiency [36], with front-surface passivation playing a crucial role. A conformal alumina film effectively reduced surface recombination through strong chemical and electrical passivation, enhancing the cell performance. Additionally, BSi cells showed a 3% increase in daily energy production due to improved angular light acceptance, highlighting their industrial potential [36].

Figure 8.6(a) shows minority carrier lifetimes of ~1 ms in BSi samples, corresponding to S_{eff} of ~20 cm s^{-1} (figure 8.6(b)). These values are slightly higher than previous reports, likely due to etching defects from the BSi process [13, 36]. Nevertheless, they remain suitable for solar applications. In contrast, the reference samples exhibit lower surface recombination velocities of ~3 cm s^{-1}. Figure 8.6(c) presents the current density–voltage (*J–V*) and power–voltage (*P–V*) characteristics of the best BSi solar cell with 80% emitter coverage, achieving > 22.1% efficiency, independently verified by Fraunhofer ISE. This represents a ~4% absolute improvement over previous BSi records. The high J_{sc} (42.2 mA cm^{-2}) and V_{oc} (665 mV) indicate excellent front-surface passivation.

The first monolithic perovskite/silicon tandem cell, with front-side BSi nanotextures and TOPCon, has been reported with improved passivation, light trapping, and perovskite wetting [53]. The tandem cell achieves a certified 28.2% efficiency, among the highest for perovskite/TOPCon and double-side-textured tandem cells.

Figure 8.7(a) presents optical photograph of silicon wafers (210 mm^2) with planar, reconstructed BSi, and pyramidal textures, highlighting the strong AR properties and scalability of reconstructed BSi. The QSSPC analysis of n-TOPCon on BSi showed that the planar sample had a high τ_{eff} of 2.6 ms, confirming strong passivation. Pristine BSi exhibited a sharp drop to τ_{eff} of 0.5 ms due to an

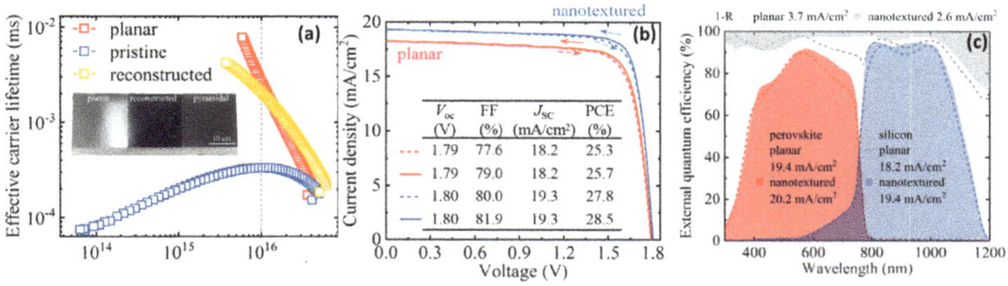

Figure 8.7. (a) Effective carrier lifetime. The inset shows a photograph showing 210 mm^2 single-crystalline silicon wafers with a planar surface (left), reconstructed BSi (middle), and pyramidal texture (right). (b) The current density–voltage (*J–V*) characteristics of the best-performing planar and nanotextured perovskite/silicon tandem solar cells, with the inset table displaying the corresponding performance parameters. (c) EQE curves for the planar and nanotextured perovskite/silicon tandem solar cells, along with the corresponding total absorbance curves (gray shaded area), 1 – *R*, where *R* represents the device reflectance. (Reproduced with permission from [53]. Copyright 2022 Elsevier.)

inhomogeneous SiO_x interlayer, while reconstructed BSi improved to τ_{eff} of 2.0 ms, indicating enhanced passivation [53].

Figure 8.7(b) shows the *J–V* curves of the best-performing nanotextured and planar tandem solar cells. The corresponding performance parameters are summarized in the inset. The nanotextured tandem cell demonstrates an outstanding PCE of 28.5% under reverse scan, with V_{oc} of 1.80 V, a current density J_{sc} of 19.3 mA cm^{-2}, and a FF of 81.9%. The device exhibits minimal hysteresis between reverse and forward scans [53].

Figure 8.7(c) highlights EQE and total absorbance measurements, demonstrating the optical benefits of transitioning from a planar to a nanotextured configuration. The planar design results in a 3.7 mA cm^{-2} current loss due to reflections, similar to previous studies [54]. The nanotexture reduces this loss to 2.6 mA cm^{-2}, close to the 2.3 mA cm^{-2} reported [55] for pyramidally textured tandems. Although reflectance increases below 600 nm, the nanotextured tandem maintains low reflectance in the 600–850 nm range, improving EQE responses for both perovskite and silicon cells due to the gradient refractive index effect. These findings demonstrate that TOPCon and surface reconstruction enable high-quality passivation while preserving light trapping, making BSi promising for the fabrication of tandem cell applications [53].

In a related study, various BSi surfaces were created using two different catalytic metals (Au and Ag) and two distinct passivation schemes for silicon solar cells [56]. Additionally, there have been other efforts in the literature. These include the following: various chemical surface treatments (alkaline etch and MACE), plasma annealing (NH_3 and H_2), and Si-rich SiN_x deposited as both a passivation and ARC. A p–n junction was formed via H_3PO_4 spraying, followed by double-layer (pm-Si/SiN_x) passivation [56]. Figure 8.8(a) presents the *I–V* characteristics of a high-performance silicon solar cell (~1 cm^2) with Ag texturing and various passivation schemes. The best results, with J_{sc} of around 40 mA cm^{-2}, were achieved using a two-step process: NH_3 plasma annealing followed by the deposition of 5 nm of pm-Si. Subsequently, a 76 nm Si-rich SiN_x layer was deposited, leading to an overall

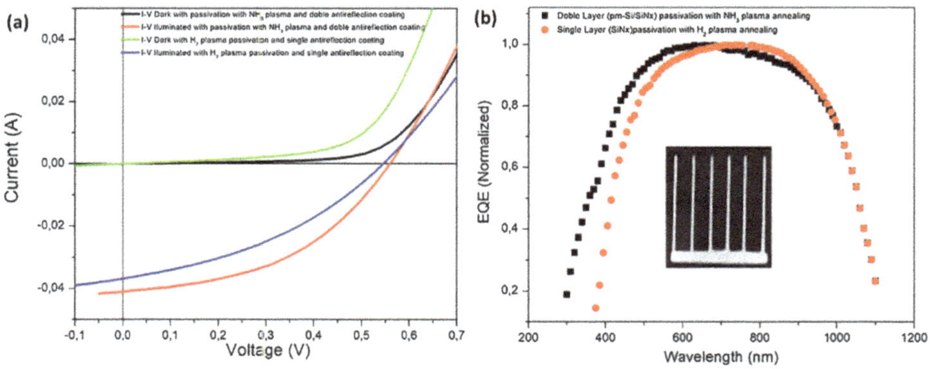

Figure 8.8. (a) Current–voltage (*I–V*) characteristics of the highest-performing silicon solar cell with silver-textured metal and various passivation schemes. (b) EQE for various passivation schemes. Inset: device photograph. (Reproduced with permission from [56]. Copyright 2019 Elsevier.)

increase in J_{sc} [57]. Figure 8.8(b) illustrates the EQE response of Ag-catalyzed solar cells with different passivation schemes. The double-stack (pm-Si/SiN_x) passivation layer with NH_3 plasma treatment outperformed conventional Si-rich SiN_x coatings with hydrogen plasma treatment [57]. *I–V* and EQE measurements confirmed enhanced I_{sc} and photon conversion efficiency, attributed to improved surface texturing, reduced optical reflectance, a downshift conversion effect, and increased EQE in the short-wavelength region [56].

8.6.2 Sensors and detectors

The sensitivity of electrical properties of BSi to environmental changes makes it ideal for sensors and photodetectors. Its high surface area enhances interaction with analytes, improving sensor performance.

The photodetection properties of sulfur-hyperdoped silicon, fabricated using a femtosecond laser, were studied across temperatures from 10 K to 300 K [58]. Hall measurements (figure 8.9(a)) show that, as temperature increases from 50 K to 300 K, more electrons are thermally excited, raising the carrier concentration. However, the concentration remains within the 10^{18} cm^{-3} range, indicating that intrinsic excitation is not dominant at 300 K. Instead, impurity ionization is the primary source of increased carriers, enhancing the photoresponse. Carrier mobility in doped silicon is influenced by ionized impurity scattering at low temperatures ($\mu_i \propto T^{3/2}$) and lattice scattering at higher temperatures ($\mu_s \propto T^{-3/2}$) [60]. Hall measurements (figure 8.9(a)) show two temperature-dependent trends consistent with these mechanisms. Mobility initially increases at low temperatures but decreases above 150 K–200 K due to intensified carrier collisions and increased lattice scattering [58].

The temperature-dependent photocurrent response of BSi detector was compared to a commercial silicon photodiode (Thorlabs FDS1010) under 900 nm illumination (figure 8.9(b)). Operating at 5 V reverse bias, the photocurrent of BSi detector was 30 times higher at room temperature, 70 times at 10 K, and 140 times at 150 K, demonstrating its superior adaptability across various temperatures [58].

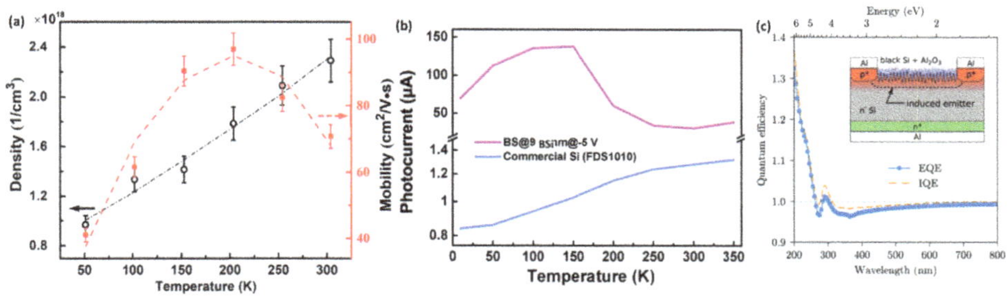

Figure 8.9. (a) Hall effect measurements of carrier concentration and mobility in the BSi sample across various temperatures. (b) Experimental comparison of photocurrent responses between the BSi photodetector and a commercial silicon-based photodiode across temperatures ranging from 10 K to 350 K under 900 nm illumination wavelength. (c) The quantum efficiency of an induced-junction BSi photodiode, external (blue circles) and internal (yellow dashed) quantum efficiency. The inset provides a schematic representation of the device structure. ((a) and (b) Reproduced with permission from [58]. Copyright 2021 Elsevier. (c) Reproduced with permission from [59]. Copyright 2020 American Physical Society.)

Therefore, sulfur-hyperdoped silicon, induced by ultrafast laser interactions, maintains high carrier activity across sub-bandgap levels, enabling a broad operational range for multifunctional optoelectronic devices. Its temperature-dependent conductivity and photoelectric performance are governed by interactions between carrier concentration and mobility.

Current UV sensors have limited performance despite their widespread use. The study in [59] investigates the potential of a BSi induced-junction photodiode for UV detection. Figure 8.9(c) presents the measured EQE and IQE of an induced-junction BSi photodiode at zero bias. The inset illustrates the schematic structure of the device. The results demonstrate EQE above unity in the photodiode, reaching over 130% at 200 nm without external amplification and exceeding 100% for wavelengths below 310 nm, significantly surpassing previous reports. The nearly identical EQE and IQE indicate minimal reflectance (~2% in UV). High sensitivity is attributed to (i) reduced Auger recombination, (ii) strong surface electric field for rapid carrier collection, (iii) nanotexturing for low reflection losses, and (iv) excellent passivation with Al_2O_3. These factors enable near 100% collection efficiency even without carrier multiplication [59].

Numerical simulations, comparisons of IQE, and spectral analysis confirm the role of BSi in enhancing the device performance, enabling above-unity efficiency at longer wavelengths and in lower bandgap materials, with major implications for UV sensors and surpassing the Shockley–Queisser limit.

8.7 Summary

This chapter discusses the unique electrical properties of black silicon (BSi), highlighting recombination mechanisms, characterization methods, passivation strategies, and optoelectronic applications. It emphasizes the importance of controlled fabrication and surface passivation for optimizing the potential of BSi in advanced technologies.

References

[1] Liu X *et al* 2014 Black silicon: fabrication methods, properties and solar energy applications *Energy Environ. Sci.* 7 3223–63

[2] Soueiti J *et al* 2023 A review of cost-effective black silicon fabrication techniques and applications *Nanoscale* **15** 4738–61

[3] Ravindra N M, Marthi S R and Sekhri S 2016 *Black Silicon: Processing, Properties, and Applications* (New York: Momentum)

[4] Otto M *et al* 2012 Extremely low surface recombination velocities in black silicon passivated by atomic layer deposition *Appl. Phys. Lett.* **100** 191603

[5] Yoo J, Yu G and Yi J 2009 Black surface structures for crystalline silicon solar cells *Mater. Sci. Eng.* B **159–160** 333–7

[6] Zhao J *et al* 1996 24% efficient perl silicon solar cell: recent improvements in high efficiency silicon cell research *Sol. Energy Mater. Sol. Cells* **41–42** 87–99

[7] Toor F *et al* 2016 Metal assisted catalyzed etched (MACE) black Si: optics and device physics *Nanoscale* **8** 15448–66

[8] von Gastrow G *et al* 2017 Recombination processes in passivated boron-implanted black silicon emitters *J. Appl. Phys.* **121** 185706

[9] Honsberg C and Bowden S 2019 Types of recombination *Photovoltaics Education* https://www.pveducation.org/pvcdrom/pn-junctions/types-of-recombination

[10] Wuerfel P 2005 Physics of solar cells: from principles to new concepts *Physics of Solar Cells: From Principles to New Concepts* ed P Würfel (New York: Wiley) p 198

[11] Ščajev P *et al* 2016 Light-induced reflectivity transients in black-Si nanoneedles *Sol. Energy Mater. Sol. Cells* **144** 221–7

[12] Liu Y *et al* 2012 Nanostructure formation and passivation of large-area black silicon for solar cell applications *Small* **8** 1392–7

[13] Repo P *et al* 2012 Effective passivation of black silicon surfaces by atomic layer deposition *IEEE J. Photovolt.* **3** 90–4

[14] Otto M *et al* 2013 Passivation of optically black silicon by atomic layer deposited Al_2O_3 *Energy Procedia* **38** 862–5

[15] Xiao G *et al* 2014 The study of defect removal etching of black silicon for solar cells *Mater. Sci. Semicond. Process.* **22** 64–8

[16] Su G *et al* 2018 The study of the defect removal etching of black silicon for diamond wire sawn multi-crystalline silicon solar cells *Sol. Energy* **170** 95–101

[17] Oh J, Yuan H-C and Branz H M 2012 An 18.2%-efficient black-silicon solar cell achieved through control of carrier recombination in nanostructures *Nat. Nanotechnol.* **7** 743–8

[18] Zhong S *et al* 2012 The study on the properties of black multicrystalline silicon solar cell varying with the diffusion temperature *Energy Procedia* **14** 505–11

[19] Shen Z *et al* 2013 Black silicon on emitter diminishes the lateral electric field and enhances the blue response of a solar cell by optimizing depletion region uniformity *Scr. Mater.* **68** 199–202

[20] Wang Y *et al* 2013 Selective nano-emitter fabricated by silver assisted chemical etch-back for multicrystalline solar cells *RSC Adv.* **3** 15483–9

[21] Kimmerle A *et al* 2016 Precise parameterization of the recombination velocity at passivated phosphorus doped surfaces *J. Appl. Phys.* **119** 025706

[22] Cuevas A *et al* 1997 The recombination velocity of boron diffused silicon surfaces *Proc. of the 14th European Photovoltaic Solar Energy Conf.*

[23] Pichler P 1996 *ICECREM Version 4.3* (*FhG-AIS Erlangen*)

[24] Yu X-Y, Li C-H and Zhao J-H 2017 Study on optical and electrical properties of gold-doped silicon fabricated by femtosecond laser *Opt. Quantum Electron.* **49** 247

[25] Zhengxi C, Yongping C and Bin M 2012 Electrical properties of black silicon *Proc. SPIE* **8419** 841918

[26] Pauw L 1958 A method of measuring specific resistivity and Hall effect of discs of arbitrary shape *Philips Res. Rep.* **13** 1–9

[27] Sze S M, Li Y and Ng K K 2021 *Physics of Semiconductor Devices* (New York: Wiley)

[28] Bscheid C *et al* 2019 Minority carrier lifetime measurements for contactless oxidation process characterization and furnace profiling *Materials* **12** 190

[29] Berger B *et al* 2011 Contactless electrical defect characterization in semiconductors by microwave detected photo induced current transient spectroscopy (MD-PICTS) and microwave detected photoconductivity (MDP) *Phys. Status Solidi* A **208** 769–76

[30] Goodarzi M *et al* 2017 A comparison between quasi-steady state and transient photoconductance lifetimes in silicon ingots: simulations and measurements *2017 IEEE 44th Photovoltaic Specialist Conf. (PVSC)*

[31] Schmidt J and Aberle A G 1997 Accurate method for the determination of bulk minority-carrier lifetimes of mono- and multicrystalline silicon wafers *J. Appl. Phys.* **81** 6186–99

[32] Parola S *et al* 2015 Time-resolved photoluminescence for the measurement of the effective carrier lifetime in Si photonic structures *Energy Procedia* **84** 25–33

[33] Schüler N *et al* 2012 Limitations in the accuracy of photoconductance-based lifetime measurements *Sol. Energy Mater. Sol. Cells* **98** 245–52

[34] Brody J, Rohatgi A and Ristow A 2003 Review and comparison of equations relating bulk lifetime and surface recombination velocity to effective lifetime measured under flash lamp illumination *Sol. Energy Mater. Sol. Cells* **77** 293–301

[35] Richter A *et al* 2012 Improved quantitative description of Auger recombination in crystalline silicon *Phys. Rev.* B **86** 165202

[36] Savin H *et al* 2015 Black silicon solar cells with interdigitated back-contacts achieve 22.1% efficiency *Nat. Nanotechnol.* **10** 624–8

[37] Muduli S P and Kale P 2023 State-of-the-art passivation strategies of c-Si for photovoltaic applications: a review *Mater. Sci. Semicond. Process.* **154** 107202

[38] Özkol E *et al* 2020 Effective passivation of black silicon surfaces via plasma-enhanced chemical vapor deposition grown conformal hydrogenated amorphous silicon layer *Phys. Status Solidi—Rapid Res. Lett.* **14** 1900087

[39] Jiang Y *et al* 2018 Improved passivation effect in multicrystalline black silicon by chemical solution pre-treatment *Appl. Phys.* A **124** 341

[40] Wang S *et al* 2021 Field-effect passivation of undiffused black silicon surfaces *IEEE J. Photovolt.* **11** 897–907

[41] Otto M *et al* 2012 Conformal Al_2O_3 coatings on black silicon by thermal ALD for surface passivation *Energy Procedia* **27** 361–4

[42] Kroll M *et al* 2010 Optical modeling of needle like silicon surfaces produced by an ICP-RIE process *Photonics for Solar Energy Systems III* (Bellingham, WA: SPIE)

[43] Toor F *et al* 2011 Multi-scale surface texture to improve blue response of nanoporous black silicon solar cells *Appl. Phys. Lett.* **99** 103501

[44] Khardani M, Bouaïcha M and Bessaïs B 2007 Bruggeman effective medium approach for modelling optical properties of porous silicon: comparison with experiment *Phys. Status Solidi* C **4** 1986–90
[45] Xie W Q, Oh J I and Shen W Z 2011 Realization of effective light trapping and omnidirectional antireflection in smooth surface silicon nanowire arrays *Nanotechnology* **22** 065704
[46] Zhang P *et al* 2014 Near-infrared optical absorption enhanced in black silicon via Ag nanoparticle-induced localized surface plasmon *Nanoscale Res. Lett.* **9** 519
[47] Füchsel K *et al* 2011 Transparent conductive oxides for nano-SIS solar cells *Proc. SPIE* **8065** 80650B
[48] Deraoui A *et al* 2013 Black silicon: microfabrication techniques and characterization for solar cells applications *Int. J. Energy Sci.* **3** 403
[49] Zhong S *et al* 2013 Influence of the texturing structure on the properties of black silicon solar cell *Sol. Energy Mater. Sol. Cells* **108** 200–4
[50] Marthi S R, Sekhri S and Ravindra N M 2015 Optical properties of black silicon: an analysis *JOM* **67** 2154–9
[51] Liu C *et al* 2021 High-efficiency black silicon tunnel oxide passivated contact solar cells achieved by adjusting the boron diffusion process *J. Mater. Sci., Mater. Electron.* **32** 23465–71
[52] von Gastrow G *et al* 2015 Analysis of the atomic layer deposited Al_2O_3 field-effect passivation in black silicon *Sol. Energy Mater. Sol. Cells* **142** 29–33
[53] Ying Z *et al* 2022 Monolithic perovskite/black-silicon tandems based on tunnel oxide passivated contacts *Joule* **6** 2644–61
[54] Bush K A *et al* 2017 23.6%-efficient monolithic perovskite/silicon tandem solar cells with improved stability *Nat. Energy* **2** 1–7
[55] Chen B *et al* 2020 Blade-coated perovskites on textured silicon for 26%-efficient monolithic perovskite/silicon tandem solar cells *Joule* **4** 850–64
[56] Ramos C *et al* 2019 A new passivation scheme for the performance enhancement of black silicon solar cells *Mater. Today Commun.* **20** 100556
[57] Mon-Pérez E *et al* 2017 Double stack layer structure of SiN_x/pm-Si thin films for downshifting and antireflection properties *Mater. Lett.* **203** 50–3
[58] Jin X *et al* 2021 High-performance black silicon photodetectors operating over a wide temperature range *Opt. Mater.* **113** 110874
[59] Garin M *et al* 2020 Black-silicon ultraviolet photodiodes achieve external quantum efficiency above 130% *Phys. Rev. Lett.* **125** 117702
[60] Zhong M *et al* 2017 High-performance photodetectors based on Sb_2S_3 nanowires: wavelength dependence and wide temperature range utilization *Nanoscale* **9** 12364–71

IOP Publishing

Black Silicon

Technology, opportunities and challenges

N M Ravindra and Priyanka Singh

Chapter 9

Material and device applications

A brief overview of the material and device applications of black silicon (BSi) is presented in this chapter. Some of the research and development (R&D) efforts are summarized. Patents and the intellectual property being pursued in BSi by various research institutions, semiconductor industries, and national labs are presented.

9.1 Introduction

In 1995, Jensen and collaborators described 'The black silicon method: a universal method for determining the parameter setting of a fluorine-based reactive ion etcher in deep silicon trench etching with profile control' [1]. Since then, the publication of the discovery of black silicon (BSi) by Eric Mazur and his team at Harvard University [2, 3], 26 years ago, has revolutionized the growth of research in BSi. Due to its inherent blackness and the benefits of utilizing silicon, and its ability to absorb 96%–98% of the incident light, BSi offers many advantages in several niche areas of the industry [4–6]. The BSi technology, developed at Harvard, has since been licensed to SiOnyx for the development and commercialization of night vision technology [7–9].

Some of the applications of BSi include the following: antibacterial surfaces, antireflection coatings, chemical and biological sensors, gas sensors, gettering of metallic impurities, infrared imagers, lab-on-a-chip, micro-electromechanical systems (MEMS), microfluidics, near-infrared and ultraviolet photodetectors, solar cells, surface-enhanced Raman spectroscopy (SERS), and terahertz spectroscopy. In general, many of these applications can be attributed to the significantly high light absorption and enhanced surface area of BSi.

Some of the R&D efforts, products, and opportunities that are being pursued utilizing BSi are described in detail in chapter 10.

doi:10.1088/978-0-7503-3247-7ch9

9.2 Applications

Table 9.1 presents examples of some of the leading BSi product manufacturers/developers/researchers in the industry today [9–23].

Table 9.1. Examples of some of the leading industries/R&D labs in BSi-related product research/manufacturing [9–23].

Device/application	Company/R&D Lab	Website; details
Black silicon masks, imaging spectrometers	NASA—Jet Propulsion Lab@ Caltech	https://microdevices.jpl.nasa.gov/capabilities/advanced-microfabrication-technologies/black-silicon/ JPL's Microdevices Laboratory has been instrumental in the development of BSi for optical applications. NASA-JPL's work focuses on creating ultraprecise optical components with reduced stray light, enhancing the performance of imaging spectrometers that are used in Earth science and space exploration [10].
Scalable production of BSi solar cells	Natcore Technology Inc.	In October 2012, scientists at Natcore Technologies reported a significant milestone by fabricating the world's first scalable BSi solar cell production process utilizing their liquid phase deposition (LPD) process [11]. This advancement held promise for enhancing solar cell efficiency and reducing manufacturing costs, especially in its approach to the fabrication of foils cells [12, 13], while potentially eliminating the use of silver [14]. Some of these efforts have focused on the fabrication of heterojunction solar cells [15, 16].
Photodetectors and imagers	SiOnyx–Harvard Team	https://www.sionyx.com SiOnyx, in collaboration with Harvard University, has been at the forefront of commercializing BSi technology. Their research focuses on developing photodetectors and imaging systems with enhanced sensitivity across a broader spectrum, including infrared wavelengths for applications in night vision, etc [9].
Solar cells, BSi nanocatalytic wet-chemical etching	National Renewable Energy Laboratory (NREL)	https://www.nrel.gov/ In 2012, a research team at NREL demonstrated an 18.2%-efficient BSi solar cell by controlling carrier recombination in nanostructures [17].

		NREL has developed a cost-effective one-step nanocatalytic wet-chemical etch that makes high efficiency solar cells [18]. One of the patents issued to the NREL team, US 11 251 318 B2, focuses on 'Efficient black silicon photovoltaic devices with enhanced blue response' [19].
BSi on highly curved surfaces	Friedrich Schiller University–Fraunhofer Insitut	In a recent report, the Institute of Applied Physics, Friedrich Schiller University in collaboration with Fraunhofer Institut for Applied Optics and Precision Engineering, has reported the 'Investigation on black silicon nanostructures fabricated by reactive ion etching on highly curved surfaces' [20].
Photodiodes: FDBS11, FDBS22, FDBS25, SM05PD8A	Thor Labs	https://www.thorlabs.com [21] The photodiodes, FDBS11, FDBS22, and FDBS25, feature a higher responsivity across the 240–1170 nm wavelength than standard photodiodes. SM05PD8A photodiodes are large area mounted BSi photodiodes.
BSi UV photodiodes, solar cells	Aalto University	https://www.aalto.fi/en/people/hele-savin Professor Hele Savin and her team at Aalto University have been recognized for their significant contributions to BSi research, including the development of solar cells and ultraviolet photodiodes. Their work involves creating high-efficiency solar cells using nanostructured BSi, aiming to improve energy conversion rates and the overall performance of photovoltaic devices [22]. Garin *et al* have reported their research on BSi ultraviolet photodiodes with external quantum efficiency above 130% [23].

It is anticipated that solar energy applications, semiconductor industries, and photonics, including photodetectors, will be the primary drivers for the expansion of BSi-related product manufacturing. SiOnyx continues to be the world leader in the manufacture of image sensors, particularly highly sensitive infrared cameras for a variety of applications. Many opportunities remain for the utilization of BSi technology, particularly in photovoltaics, and civilian and military applications.

9.3 Historic perspective

Examples of the ability to manipulate the incident light–material interaction for improved device performance via (a) modification of material surface morphology by methods such as sandblasting, (b) utilizing reflection at material surfaces/interfaces, or (c) having sloped features are illustrated in table 9.2. Forbes presents an excellent history of BSi in photodetector patents [24].

Table 9.2. Examples of patents that have utilized various methods for the manipulation of light–material interactions.

Patent/date/assignee	Abstract
US 3 487 223 30 December 1969 Multiple internal reflection structure in a silicon detector which is obtained by sandblasting, United States Department of the Air Force	A silicon photoconductor apparatus having improved efficiency in infrared radiation detection by sandblasting the reflective surface of the apparatus and thereby creating a near-infinite number of reflecting angles and a corresponding multiplication of the quantum energy yield.
US 4 419 533 6 December 1983 Photovoltaic device having incident radiation directing means for total internal reflection, Energy Conversion Devices, Inc.	There is disclosed new and improved photovoltaic devices which provide increased short circuit currents and efficiencies over that previously obtainable from prior devices. The disclosed devices include incident radiation directing means for directing at least a portion of the incident light through the active region or regions thereof at angles sufficient to substantially confine the directed radiation in the devices. This allows substantially total utilization of photogenerated electron–hole pairs. Further, because the light is directed through the active region or regions at such angles, the active regions can be made thinner to also increase collection efficiencies. The incident radiation directors can be random surface or bulk reflectors to provide random scattering of the light, or periodic surface or bulk reflector to provide selective scattering of the light. While the present invention is applicable to photovoltaic devices formed from any type of semiconductor material, as for example, crystalline, polycrystalline, or amorphous semiconductor alloys or any combination thereof, disclosure herein is primarily directed to

	photovoltaic devices formed from amorphous silicon alloys preferably incorporating fluorine as a density of states reducing element. The disclosure is also directed to, without limitation, photovoltaic devices of the p–i–n configuration, both as single cells and multiple cells arranged in tandem.
US 7 456 452 B2 25 November 2008 Light sensor having undulating features for CMOS imager, Micron Technology, Inc.	Light sensors in an imager having sloped features including, but not limited to, hemispherical, v-shaped, or other sloped shapes. Light sensors having such a sloped feature can redirect incident light that is not absorbed by one portion of the photosensor to another portion of the photosensor for absorption there.

9.4 Patents/invention disclosures

In table 9.3, a summary of the patents/invention disclosures of BSi is presented.

Table 9.3. Summary of patents/invention disclosures on BSi.

Patent number/patent application publication number Date Patent/application title current assignee	Abstract
US 10 741 399 B2 11 August 2020 Femtosecond laser-induced formation of submicrometer spikes on a semiconductor substrate, President and Fellows of Harvard College	The present invention generally provides semiconductor substrates having submicron-sized surface features generated by irradiating the surface with ultra short laser pulses. In one aspect a method of processing a semiconductor substrate is disclosed that includes placing at least a portion of a surface of the substrate in contact with a fluid, and exposing that surface portion to one or more femtosecond pulses so as to modify the topography of that portion. The modification can include, e.g. generating a plurality of submicron-sized spikes in an upper layer of the surface.

(*Continued*)

Table 9.3. (*Continued*)

Patent number/patent application publication number Date Patent/application title current assignee	Abstract
US2007/06167011 5 December 2006 Etching method, Oki Electric Industry Co., Ltd	In an etching method of the present invention, a first etching step for etching a silicon semiconductor region at a first etching rate with use of a first etching gas is first performed. Then, after the first etching step, a second etching step is performed in order to etch the silicon semiconductor region at a second etching rate that is lower than the first etching rate by using a second etching gas that includes carbon and fluoride, with the ratio of the fluoride in the second etching gas being higher than that of the carbon therein. Carbon is included in the second etching gas in order to chemically bind to at least either oxygen or hydrogen used in the first etching step. Therefore, it is possible to inhibit the generation of black silicon (i.e. a residue composed of silicon needles) which is caused as a result of attachment of oxygen and hydrogen on the etching surface. The resulting etched surface is smoothly formed, and black silicon is not formed thereon.
W02002029858A2 30 August 2001 Deep trench etching method to reduce/eliminate formation of black silicon, Infineon Technologies North America Corp.	In a method of etching a wafer to form a DT (deep trench) in a plasma reactor, wherein the wafer temperature is greater than the cathode temperature, the improvement of conducting etching to reduce or eliminate 'black silicon', comprising: (a) providing a plasma etch reactor comprising walls defining an etch chamber; (b) providing a plasma source chamber remote from and in communication with the etch chamber, and a wafer chuck or pedestal disposed in the etch chamber to seat a wafer; (c) forming a plasma within the plasma source chamber and providing the plasma to the etch chamber; (d) supplying RF energy to etch the wafer to a point where the wafer temperature is greater than the cathode temperature; and (e) increasing the

	flow rate of fluorine species near the end of the DT process to provide the isotropic component needed to widen the CD.
US 9 276 143 B2 1 March 2016 Silicon-based visible and near-infrared optoelectronic devices, President and Fellows of Harvard College	In one aspect, the present invention provides a silicon photodetector having a surface layer that is doped with sulfur inclusions with an average concentration in a range of about 0.5 atom percent to about 1.5 atom percent. The surface layer forms a diode junction with an underlying portion of the substrate. A plurality of electrical contacts allow application of a reverse bias voltage to the junction in order to facilitate generation of an electrical signal, e.g., a photocurrent, in response to irradiation of the surface layer. The photodetector exhibits a responsivity greater than about 1 A/W for incident wavelengths in a range of about 250 nm to about 1050 nm, and a responsivity greater than about 0.1 A/W for longer wavelengths, e.g., up to about 3.5 microns.
W02001099159A2 20 June 2001 Reduction of black silicon in deep trench etch, Infineon Technologies North America Corp.	In a method of etching a wafer in a plasma etch reactor, the improvement of conducting etching to reduce or eliminate 'black silicon' comprising: (a) providing a plasma etch reactor comprising walls defining an etch chamber; (b) providing a plasma source chamber remote from and in communication with the etch chamber to provide a plasma to the etch chamber, and a wafer chuck or pedestal disposed in the etch chamber to seat a wafer; (c) providing a dielectric wall in proximity to and around a periphery of the wafer; (d) providing a modification to a lower RF electrode by interposing conductor means into an extension of Vdc flat sheath boundary relationship to the dielectric wall means and the wafer or in substitution for the dielectric wall; (e) forming a plasma within the plasma source chamber and providing the plasma to the etch chamber; and (f) supplying RF energy to the wafer chuck to assist etching of the wafer by forming electric fields between the upper surface of the wafer and

(*Continued*)

Table 9.3. (*Continued*)

Patent number/patent application publication number Date Patent/application title current assignee	Abstract
	the walls of the etch chamber, to provide extension of a Vdc flat sheath boundary beyond and into a defocusing relationship to the wafer edge to reduce mask erosion and eliminate occurrence of 'black silicon' formation.
US 7 291 286 B2 6 November 2007 Methods for removing black silicon and black silicon carbide from surfaces of silicon and silicon carbide electrodes for plasma processing apparatuses, Lam Research Corporation	Methods for removing black silicon or black silicon carbide from a plasma-exposed surface of an upper electrode of a plasma processing chamber are provided. The methods include forming a plasma using a gas composition containing a fluorine-containing gas, and removing the black silicon or black silicon carbide from the surface with the plasma. The methods can also remove black silicon or black silicon carbide from surfaces of the components in the chamber in addition to the upper electrode.
US2013/0068297 A1* 21 March 2013 Black silicon solar cell and its preparation method, The Institute of Microelectronics of Chinese Academy of Sciences	A black silicon solar cell includes a metal back electrode, the crystal silicon, a black silicon layer, a passivation layer and a metal gate; wherein, the metal back electrode is located on the back surface of the crystal silicon, the black silicon layer is located on the crystal silicon, the passivation layer is located on the black silicon layer, the metal gate is located on the passivation layer. The fabrication method includes: carrying out pretreatment of the silicon wafer; preparing the black silicon layer on the surface of the pretreated silicon wafer by using plasma immersion ion implantation technology; preparing an emitter on the black silicon layer, and carrying out passivation treatment on the emitter to form the passivation layer; respectively preparing the metal back electrode and the metal gate on the back surface of the single crystal silicon wafer and the passivation layer, respectively.

CN10191678A** 8 February 2012 A kind of black silicon solar cell and preparation method thereof, The Institute of Microelectronics of Chinese Academy of Sciences	The invention discloses a black silicon solar cell and a preparation method thereof, and belongs to the technical field of solar cell device manufacturing. The black silicon solar cell comprises a metal back electrode, crystalline silicon, a black silicon layer, a passivation layer and a metal grid; the metal back electrode is positioned on the back surface of the crystalline silicon, the black silicon layer is positioned on the crystalline silicon, the passivation layer is positioned on the black silicon layer, and the metal grid electrode is positioned on the passivation layer. The preparation method comprises the following steps: preprocessing a silicon wafer; preparing a black silicon layer on the surface of the pretreated silicon wafer by using a plasma immersion ion implantation technology; preparing an emitter on the black silicon layer, and passivating the emitter to form a passivation layer; and respectively preparing a metal back electrode and a metal grid electrode on the back surface of the monocrystalline silicon wafer and the passivation layer. The invention uses the black silicon layer as the absorption layer, and the average absorption rate of the battery is higher in the visible light wave band.
US 2019/0188471 A1 20 June 2019 Method and apparatus for Biometric data capture, Microsoft Technology Licensing, LLC	A method and apparatus for biometric data capture are provided. The apparatus includes interactive head mounted eyepiece worn by a user that includes an optical assembly through which a user views a surrounding environment and displayed content. The optical assembly comprises a corrective element that corrects the user's view of the surrounding environment and an integrated processor for handling content to the user. An integrated optical sensor captures biometric data when the eyepiece is positioned so that a nearby individual is proximate to the eyepiece. Biometric data is captured using the eyepiece and is transmitted to a remote processing facility for

(*Continued*)

Table 9.3. (*Continued*)

Patent number/patent application publication number Date Patent/application title current assignee	Abstract
	interpretation. The remote processing facility interprets the captured biometric data and generates display content based on the interpretation. This display content is delivered to the eyepiece and displayed to the user.
US 2020/0192089 A1 18 June 2020 Head-worn adaptive display, Microsoft Technology Licensing, LLC	This disclosure concerns an interactive head-mounted eyepiece with an integrated processor for handling content for display and an integrated image source for introducing the content to an optical assembly through which the user views a surrounding environment and the displayed content. The optical assembly includes absorptive polarizers or anti-reflective coatings to reduce stray light.
US 11 251 318 B2 15 February 2022 Efficient black silicon photovoltaic devices with enhanced blue response, Alliance for Sustainable Energy, LLC	A photovoltaic (PV) device with improved blue response. The PV device includes a silicon substrate with an emitter layer on a light receiving side. The emitter layer has a low dopant level such that it has sheet resistance of 90–170 ohm sq^{-1}. Anti-reflection in the PV device is provided solely by a nano-structured or black silicon surface on the light receiving surface, through which the emitter is fanned by diffusion. The nanostructures of the black silicon are formed in a manner that does not result in gold or another high recombination metal being left in the black silicon such as with metal-assisted etching using silver. The black silicon is further processed to widen these pores so as to provide larger nanostructures with lateral dimensions in the range of 65–150 nanometers so as to reduce surface area and also to etch away a highly doped portion of the emitter.

US 9 222 837 B2 29 December 2015 Black silicon-based high-performance MEMS thermopile IR detector and fabrication method, Jiangsu R&D Center for Internet of Things	This invention involves structure and fabrication methods of a black silicon-based MEMS thermopile IR detector. The high-performance black silicon-based MEMS thermopile IR detector includes a substrate; a releasing barrier band on the substrate; a thermal isolation cavity constructed by the releasing barrier band; a black silicon-based IR absorber located right above the thermal isolation cavity; a number of thermocouples are set around the lateral sides of the black silicon-based IR absorber. The thermopiles around the black silicon-based IR absorb are electrically connected in series thus to form a thermopile. Metallic electrodes are located beside the electrically connected thermopiles for signal output. The cold junctions of the thermopile are connected to the substrate through the first thermal-conductive-electrical-isolated structures, the heat conductor is located at the lateral sides of the thermal isolation cavity. The hot junctions of the thermopile are in contact with the IR absorber through the second thermal-conductive-electrical-isolated structures, and the second thermal-conductive-electrical-isolated structures are located above the releasing barrier band. The structure of such a detector is simple, and it is easy to implement and can also be monolithically integrated. Such a detector has high responsivity and detection rate, and is CMOS-compatible, thus can be used widely in a safe and reliable manner.
US 8 964 298 B2 24 February 2015 Video display modification based on sensor input for a see-through near-to-eye display, Microsoft Corporation	This disclosure concerns a near field communication (NFC) device which includes a wrist-worn NFC-enabled electronics device, wherein the wrist-worn NFC enabled electronics device includes a first communications link for communicating with a second NFC-enabled electronics device via NFC protocols, and a second communications link for communicating with an eyepiece via a medium-range communications protocol and receiving

(*Continued*)

Table 9.3. (*Continued*)

Patent number/patent application publication number Date Patent/application title current assignee	Abstract
	control commands. The wrist-worn NFC-enabled electronics device facilitates the transfer of data between the eyepiece and the second NFC-enabled electron Ices device. The eyepiece comprises optics enabling a see-through display on which is displayed the data.
US 10 837 061 B2 17 November 2020 Enrichment and detection of nucleic acids with ultra-high sensitivity, The Regents of the University of California	Methods, systems, and devices are disclosed for enrichment and detection of molecules of a target biomarker. In one aspect, In one aspect, a biosensor device for enriching and detecting biomarker molecules includes a substrate, and a microarray of hydrophilic islands disposed on the substrate. A sensing area on each of the microarray hydrophilic islands is structured to anchor bio-molecular probes of at least one type for detecting molecules of a target biomarker and to attract an array of nanodroplets of a biomarker solution that includes the target biomarker molecules. A hydrophobic surface is disposed to surround the microarray of hydrophilic islands.
US 10 860 100 B2 8 December 2020 AR glasses with predictive control of external device based on event input, Microsoft Corporation	This disclosure concerns an interactive head-mounted eye piece with an integrated processor for handling content for display and an integrated image source for introducing the content to an optical assembly through which the user views a surrounding environment and the displayed content, wherein the eyepiece includes predictive control of external device based on an event input.
US 10 852 540 B2 1 December 2020 AR glasses with event and user action control of external applications, Microsoft Technology Licensing, LLC	This disclosure concerns an interactive head-mounted eye piece with an integrated processor for handling content for display and an integrated image source for introducing the content to an optical assembly through which the user views a surrounding environment and the displayed

content, wherein the eyepiece includes event and user action control of external applications.

US 11 369 968 B2
28 June 2022
Molecular manipulation and assay with controlled temperature (II),
Essenlix Corporation

The present invention provides devices, systems, and methods for rapid and easy-to-use in sample thermal cycling or temperature changes for the facilitation of reactions such as but not limited to PCR.

US 11 323 829 B2
3 May 2022
Round window coupled hearing systems and methods,
EarLens Corporation

Support can be configured for placement in the middle ear to couple a transducer to the round window, such that the transducer can be removed from the round window without damaging the round window. The support can be configured to couple the transducer to the sound window such that the support can be removed from the round window. The support may be configured to decouple the transducer from the round window such that the transducer can be removed from the middle ear of the user, for example when the support is affixed to the middle ear. Removal of the transducer from the middle ear without damaging the round window can allow safe removal of the transducer, for example when the patient wishes to receive MRI imaging.

US 11 296 137 B2
5 April 2022
High quantum efficiency Geiger-mode avalanche diodes including high sensitivity photon mixing structures and arrays thereof,
Sense Photonics Inc.

A photodetector device includes a semiconductor material layer and at least one photodiode in the semiconductor material layer. The at least one photodiode is configured to be biased beyond a breakdown voltage thereof to generate respective electrical signals responsive to detection of incident photons. The respective electrical signals are independent of the optical power of the incident photons. A textured region is coupled to the semiconductor material layer and includes optical structures positioned to interact with the incident photons in the detection thereof by the at least one photodiode. Two or more photodiodes may define a pixel of the photodetector device, and the optical structures may be configured to direct the incident photons to any of the two or more photodiodes of the pixel.

(*Continued*)

Table 9.3. (*Continued*)

Patent number/patent application publication number Date Patent/application title current assignee	Abstract
US 11 121 166 B2 14 September 2021 Image sensor device, Taiwan Semiconductor Manufacturing Co., Ltd	An image sensor device is provided. The image sensor device includes a semiconductor substrate including a front surface, a back surface opposite the front surface, and a light-sensing region extending from the front surface into the semiconductor substrate. The image sensor device includes a light- blocking structure in the semiconductor substrate and surrounding the light-sensing region. The light-blocking structure includes a conductive light reflection structure and a light absorption structure, and the light absorption structure is between the conductive light reflection structure and the back surface. The image sensor device includes an insulating layer between the light-blocking structure and the semiconductor substrate.
US 2019/0109317 A1 11 April 2019 Nanoporous silicon and lithium ion battery anodes formed therefrom, University of Southern California	An electrode for a lithium-ion battery, the electrode including nanoporous silicon structures, each nanoporous silicon structure defining a multiplicity of pores, a binder, and a conductive substrate. The nanoporous silicon structures are mixed with the binder to form a composition, and the composition is adhered to the conductive substrate to form the electrode. The nanoporous silicon may be, for example, nanoporous silicon nanowires or nanoporous silicon formed by etching a silicon wafer, metallurgical grade silicon, silicon nanoparticles, or silicon prepared from silicon pre cursors in a plasma or chemical vapor deposition process. The nanoporous silicon structures may be coated or combined with a carbon-containing compound, such as reduced graphene oxide. The electrode has a high specific capacity (e.g., above 1000 mAh g^{-1} at current rate of 0.4 A g^{-1}, above 1000 mAh g^{-1} at a current rate of 2.0 A g^{-1}, or above 1400 mAh g^{-1} at a current rate of 1.0 A g^{-1}).

US 10 067 270 B2 4 September 2018 Electromagnetic wave absorbing/radiating material, method for manufacturing same, and infrared source, National Institute for Materials Science	The present invention relates to an electromagnetic wave absorbing/radiating material which includes: a conductor; and a plurality of conductor discs disposed in an array above the surface of the conductor or a perforated conductor layer with a plurality of holes defined in an array above the surface of the conductor.
US 10 718 014 B2 21 July 2020 Thermo-controllable high-density chips for multiplex analyses, Takaro Bio USA Inc.	The present invention provides miniaturized instruments for conducting chemical reactions where control of the reaction temperature is desired or required. Specifically, this invention provides chips and optical systems for performing and monitoring temperature-dependent chemical reactions. The apparatus and methods embodied in the present invention are particularly useful for high-throughput and low-cost amplification of nucleic acids.
US 11 085 825 B2 10 August 2021 Self-referenced spectrometer, SI Ware Systems Inc.	Aspects of the disclosure relate to a self-referenced spectrometer for providing simultaneous measurement of a back ground or reference spectral density and a sample or other spectral density. The self-referenced spectrometer includes an interferometer optically coupled to receive an input beam and to direct the input beam along a first optical path to produce a first interfering beam and a second optical path to produce a second interfering beam, where each interfering beam is produced prior to an output of the interferometer. The spectrometer further includes a detector optically coupled to simultaneously detect a first interference signal produced from the first interfering beam and a second interference signal produced from the second interfering beam, and a processor configured to process the first interference signal and the second interference signal and to utilize the second interference signal as a reference signal in processing the first interference signal.
US 10 090 161 B2 2 October 2018	A plasma etching apparatus performs plasma etching on a substrate having a resist pattern

(*Continued*)

Table 9.3. (*Continued*)

Patent number/patent application publication number Date Patent/application title current assignee	Abstract
Plasma etching method, Tokyo Electron Limited	formed thereon and an outer edge portion where the substrate surface is exposed. The plasma etching apparatus includes a support part that supports the substrate, a cover member that covers the outer edge portion of the substrate and prevents plasma from coming around the outer edge portion, and a control unit that generates plasma by controlling high frequency power application and supply of a processing gas for etching, and uses the generated plasma to etch the substrate that is supported by the support part and has the outer edge portion covered by the cover member. After etching the substrate, the control unit generates plasma by controlling high frequency power application and supply of a processing gas for ashing, and uses the generated plasma to perform ashing on the resist pattern on the etched substrate.
US 9 749 758 B2 29 August 2017 Transducer devices and methods for hearing, EarLens Corp	A device to transmit an audio signal to a user comprises a transducer and a support. The support is configured for placement on the eardrum to drive the eardrum. The transducer is coupled to the support at a first location to decrease occlusion and a second location to drive the eardrum. The transducer may comprise one or more of an electromagnetic balanced armature transducer, a piezoelectric transducer, a magnetostrictive transducer, a photostrictive transducer, or a coil and magnet. The device may find use with open canal hearing aids.
EP 3 364 759 B1 9 December 2020 Antimicrobial coatings, Agency for Science Technology and Research Singapore	Infectious diseases and the increasing threat of worldwide pandemics have underscored the importance of antibiotics and hygiene. Microbial infection is also one of the most serious concerns for many commercial applications, such as medical devices and hospital surfaces, textiles, food packaging,

children's toys, electrical appliances, and dental surgery equipment. Intensive efforts have been devoted to creating self-disinfecting and microbicide surfaces, mainly achieved by coating microbicides onto surfaces to biochemically reduce the infectivity of microbes. However, critical challenges remain at this point such as growing drug resistance to the microbicide agents, low microbial killing efficacy and poor long-term stability of the coated surfaces.

AU 2018203550 B2
21 May 2018
Optically coupled cochlear implant systems and methods,
EarLens Corp

A system (10) to stimulate tissue of a user having a tympanic membrane, the system (10) comprising: a plurality of electrodes (32E) configured for placement at least partially within the tissue; circuitry configured to receive a signal from a source; at least one light source coupled to the circuitry, the at least one light source and the circuitry configured to transmit through the tympanic membrane a multiplexed optical signal comprising a plurality of light pulses; at least one photo detector (34) sized and shaped to collect a scattered multiplexed optical signal transmitted through the tympanic membrane, the at least one photodetector (34) and the plurality of electrodes (32E) configured to pass current through the electrodes (32E) in response to the scattered multiplexed optical signal comprising the plurality of light pulses in order to stimulate the tissue, wherein the multiplexed optical signal comprises a plurality of optical channels, each optical channel of the plurality corresponding to at least one frequency of the sound, and wherein the multiplexed optical signal is transmitted through the tympanic membrane of the user to the at least one photodetector (34), the at least one photodetector (34) configured to be affixed to the middle ear and coupled to the plurality of electrodes (32e) configured to be positioned at least partially within the cochlea.

(*Continued*)

Table 9.3. (*Continued*)

Patent number/patent application publication number Date Patent/application title current assignee	Abstract
US 11 174 167 B1 16 November 2021 Silicon carbon composites comprising ultralow *Z*, Group14 Technologies Inc.	Silicon–carbon composite materials and related processes are disclosed that overcome the challenges for providing amorphous nano-sized silicon entrained within porous carbon. Compared to other, inferior materials and processes described in the prior art, the materials and processes disclosed herein find superior utility in various applications, including energy storage devices such as lithium ion batteries.
US 11 264 371 B2 1 March 2022 US 8 692 198 B2 8 April 2014 Photosensitive imaging devices and associated methods, SiOnyx LLC	A monolithic sensor for detecting infrared and visible light according to an example includes a semiconductor substrate and a semiconductor layer coupled to the semiconductor substrate. The semiconductor layer includes a device surface opposite the semiconductor substrate. A visible light photo diode is formed at the device surface. An infrared photo diode is also formed at the device surface and in proximity to the visible light photodiode. A textured region is coupled to the infrared photodiode and positioned to interact with electromagnetic radiation.
US 11 495 793 B2 8 November 2022 Composite carbon materials comprising lithium alloying electrochemical modifiers, Group14 Technologies, Inc.	The present application is generally directed to composites comprising a hard carbon material and an electrochemical modifier. The composite materials find utility in any number of electrical devices, for example, in lithium ion batteries. Methods for making the disclosed composite materials are also disclosed.
US 9 762 830 B2 12 September 2017 High dynamic range CMOS image sensor having anti-blooming properties and associated methods, SiOnyx LLC	A method of providing blooming protection to a CMOS imager having a pixel array of a plurality of pixels arranged in rows and columns, where the CMOS imager is operable to capture high dynamic range images using a rolling shutter, is provided. Such a method can include reading out charge accumulated by the pixels in a readout

row of a first integration time, applying a reset to the readout row for a reset time sufficient to allow readout and reset to occur in at least one subsequent row, and starting a second integration time of the pixels in the readout row, wherein the second integration time is shorter than the first integration time, and wherein the at least one subsequent row is a sufficient number of rows to have a combined reset to preclude blooming effects from the pixel array during the second integration time.

US 10 816 939 B1
27 October 2020
Method of illuminating an environment using an angularly varying light emitting device and an imager,
Zane Coleman

In one embodiment, a method of illuminating an environment comprises angular cycling angular light flux output from light sources in angular bins of an angularly varying light emitting device, capturing using an imager a plurality of images of the environment synchronized with the light sources, determining a light property of one or more spatial zones associated with the angular bins, and adjusting the light flux output from one or more light sources based on the light properties. In one embodiment, the light emitting device comprises a micro-LED array comprising the plurality of light sources and/ or the imager. In one embodiment, the light flux output is adjusted to achieve a target illuminance in one or more spatial zones. In one embodiment, the angular widths in theta and phi spherical coordinates of the angular bins are less than 2 degrees.

US 12 185 442 B2
31 December 2024
Method of illuminating an environment using an angularly varying light emitting devices and a target illuminance for a spatial zone,
Zane Coleman

A method for illuminating an environment using an angularly varying light emitting device including determining a target illuminance based on a specification, identifying angular bins corresponding to illuminating one or more spatial zones or areas, and adjusting a first light flux output of at least one light source into angular bins based at least in part on the target illuminance. Adjusting the first light flux output may include adjusting it such that it is different from a second light flux output into a second angular bin. An imager may be used

(*Continued*)

Table 9.3. (*Continued*)

Patent number/patent application publication number Date Patent/application title current assignee	Abstract
	to identify an environment, activity, or estimate or measure the illuminance of the environment. Adjusting the light flux output in a bin may be based on daylight or traditional light fixtures or lamps. An angularly varying light emitting device or system may adjust the light flux output in an angular bin based on a target illuminance.
US 9 927 188 B2 27 March 2018 Metamaterials-enhanced passive radiative cooling panel, Palo Alto Research Center Inc.	A low-cost passive radiative cooling panel includes an emitter layer disposed under an upper reflective layer, where the emitter layer includes metamaterial nanostructures (e.g., tapered nanopores) configured to dissipate heat in the form of radiant energy that is transmitted through the reflective layer into cold near-space. In an embodiment the emitter layer includes ultra-black material configured to emit, with an emissivity close to unity, radiant energy having wavelengths/frequencies that fall within known atmospheric transparency windows (e.g., 8–13 μm or 16–28 μm). In a practical embodiment the emitter layer is formed using a modified Anodic Aluminum Oxide (AAO) self-assembly technique followed by electroless plating that forms metal-plated tapered nanopores. The reflective layer includes a distributed Bragg reflector configured to reflect at least 94% of incident solar light while passing the emitted radiant energy, and in some embodiments is implemented using a low-cost, commercially available solar mirror film.
US 10 276 740 B1 30 April 2019 Co-deposition of black silicon, The United States of America, as represented by the Secretary of the Air Force	A method for forming a photon absorbing layer. A substrate having a target surface is introduced into a controllable environment, and the pressure within the controllable environment is reduced. A first flux of a semiconductor material and a second flux of a dopant are simultaneously directed toward

			the target surface for a period of time, thereby producing a thickness of a substantially amorphous layer of the semiconductor material and dopant on the target surface. The semiconductor layer is laser annealed to convert it to a substantially multi-crystalline layer.
US 10 446 838 B2	15 October 2019	Negative electrode for nonaqueous secondary battery and nonaqueous secondary battery, negative electrode active material and method for producing same, complex including nano silicon, carbon layer, and cationic polymer layer, and method for producing complex formed of nano silicon and carbon layer, Toyota Industries Corp	Cycle characteristics of a nonaqueous secondary battery are to be improved. An active material including: a first active material that contains a nano silicon produced by heating a layered polysilane represented by a composition formula (SiH), and having a structure in which multiple six-membered rings formed from silicon atoms are connected; and a second active material that contains a graphite, is used in a negative electrode. With this, expansion and contraction due to stress during charging and discharging can be mitigated, and thereby cycle characteristics improve.
US 10 453 976 B2	22 October 2019	Systems and methods for forming foil contact rear emitter solar cells with carrier selective contacts, Natcore Technology, Inc.	A solar cell structure may provide a front surface that may include a front passivation layer and front anti-reflective layer. The solar cell structure may provide both contacts on a rear surface. In some cases, the rear surface may optionally provide passivation, doped, and/or transparent conductive oxide layers. The rear surface also provides a multilayer foil assembly (MFA). The MFA provides a first metal foil in electrical communication with doped regions of the rear surface of the substrate, such as base or emitter regions. The MFA may also provide a second metal foil that is spaced apart from the first metal foil by a dielectric layer. The metal foils and dielectric layers may include openings through the entirety of these layers, and these openings may be utilized to form contacts electrically coupled to the second metal foil, which is electrically isolated from the first metal foil.
US 9 484 190 B2	1 November 2016		Proposed is a showerhead-cooler system of a semiconductor-processing chamber with uniform distribution of plasma density. The

(*Continued*)

Table 9.3. (*Continued*)

Patent number/patent application publication number Date Patent/application title current assignee	Abstract
Showerhead-cooler system of a semiconductor-processing chamber for semiconductor wafers of large area, Yuri Glukhoy	showerhead has a plurality of through gas holes that are coaxial with respective channels of the gas-feeding cooler plate. On the gas inlet side, the though passages of the showerhead are provided with unequal conical nozzles characterized by a central angle that decreases from the peripheral part of the showerhead to the showerhead center. Such design provides uniformity of plasma density. Furthermore, in order to protect the walls of the nozzle and the walls of the gas holes from erosion that may be caused by the hollow-cathode phenomenon, these areas are coated with a thin protective coating that is resistant to electrical break down and chemical corrosion.
US 11 121 166 B2 14 September 2021 Image sensor device, Taiwan Semiconductor Manufacturing Co TSMC Ltd	An image sensor device is provided. The image sensor device includes a semiconductor substrate including a front surface, a back surface opposite to the front surface, and a light-sensing region extending from the front surface into the semiconductor substrate. The image sensor device includes a light-blocking structure in the semiconductor substrate and surrounding the light-sensing region. The light-blocking structure includes a conductive light reflection structure and a light absorption structure, and the light absorption structure is between the conductive light reflection structure and the back surface. The image sensor device includes an insulating layer between the light-blocking structure and the semiconductor substrate.
AU 2019201444 B2 1 March 2019 A method for producing a textured structure of a crystalline silicon solar cell, CSI Cells Co Ltd	The present invention discloses a method for producing a textured structure of a crystalline silicon solar cell, comprising the following steps: (1) forming a porous layer structure on the surface of a silicon wafer; (2) then cleaning with a first alkaline chemical solution; (3) removing residual metal

	particles with a cleaning solution; (4) and then etching the surface with a first chemical etching solution to obtain the textured structure of the crystalline silicon solar cell. The present invention greatly prolongs the lifetime of the mixed solution of hydrofluoric acid and nitric acid and ensures the stability and uniformity of the textured structure.
US 11 227 886 B2 18 January 2022 Mechanisms for forming image sensor device, Taiwan Semiconductor Manufacturing Co TSMC Ltd	An image sensor device is provided. The image sensor device includes a semiconductor substrate and a light sensing region in the semiconductor substrate. The image sensor device also includes a dielectric layer over the semiconductor substrate and a filter partially surrounded by the dielectric layer. The filter has a protruding portion protruding from a bottom surface of the dielectric layer. The image sensor device further includes a shielding layer between the dielectric layer and the semiconductor substrate and surrounding the protruding portion of the filter. In addition, the image sensor device includes a reflective element between the shielding layer and an edge of the light sensing region.
US 2019/0109317 11 April 2019 Nanoporous silicon and lithium ion battery anodes formed therefrom, University of Southern California	An electrode for a lithium ion battery, the electrode including nanoporous silicon structures, each nanoporous silicon structure defining a multiplicity of pores, a binder, and a conductive substrate. The nanoporous silicon structures are mixed with the binder to form a composition, and the composition is adhered to the conductive substrate to form the electrode. The nanoporous silicon may be, for example, nanoporous silicon nanowires or nanoporous silicon formed by etching a silicon wafer, metallurgical grade silicon, silicon nanoparticles, or silicon prepared from silicon precursors in a plasma or chemical vapor deposition process. The nanoporous silicon structures may be coated or combined with a carbon-containing compound, such as reduced graphene oxide. The electrode has a high specific capacity (e.g., above 1000 mAh g^{-1} at current rate of

(*Continued*)

Table 9.3. (*Continued*)

Patent number/patent application publication number Date Patent/application title current assignee	Abstract
	0.4 A g^{-1}, above 1000 mAh g^{-1} at a current rate of 2.0 A g^{-1}, or above 1400 mAh g^{-1} at a current rate of 1.0 A g^{-1}).
US 2019/0245155 A1 8 August 2019 Methods, products, and systems relating to making, providing, and using nano-crystalline cellulose superlattice solar cells to produce electricity, Stephan Heath	Nanocrystalline superlattice solar cells including nanocrystalline cellulose (NCC) that overcomes one or more of the problems existing in the art are provided including nanocrystalline superlattice solar cells that will add one or more layers to the nanocrystalline solar cell to produce electricity that can be used for solar energy devices, solar energy storage systems, applications, products and services utilizing NCC in the construction of one or more of the nanocrystalline superlattice layers of the solar cell.
CN109148492B 25 November 2022 Light sensing device, Taiwan Semiconductor Manufacturing Co TSMC Ltd	A structure and a manufacturing method of a photo sensor device are provided. The light sensing device comprises a semiconductor substrate and a light sensing area in the semiconductor substrate. The light sensing device also includes a filter element over the semiconductor substrate and aligned with the light sensing region. The filter element is provided with a first part and a second part, and the first part is positioned between the second part and the light sensing area. The light sensing device also includes a light blocking element over the semiconductor substrate and beside the first portion of the filter element. Furthermore, the light sensing device comprises a dielectric element on the shading element and beside the second portion of the filter element. The top width of the shading element is larger than the bottom width of the dielectric element.
US 9 312 517 B2 12 April 2016 Electronic device displays with border masking layers,	An electronic device may be provided with a display. The display may have an active area with an array of display pixels and an inactive border region containing metal lines

Apple Inc.	and other Support circuitry. The array of display pixels may be formed from organic light-emitting diode structures in an organic light-emitting diode layer. The display may also include an encapsulant layer on the organic light-emitting diode layer and a substrate layer. An opaque masking layer may be formed in the inactive border region. The opaque masking layer may overlap structures in the inactive border region Such as the Support circuitry, thereby blocking the Support circuitry from view. The opaque masking layer may be formed a glass layer in the display, in part of a polarizer in a display, under a planarization layer, or between other layers in the display.
US 9 837 561 B2* 5 December 2017 Laser processed back contact heterojunction solar cells, Natcore Technology Inc.	An interdigitated solar cell may provide a heterojunction or tunnel junction emitter and base contacts that comprise laser processed regions that electrically couple the base contact to a substrate. Methods for manufacturing such solar cells to provide interdigitated back contacts may utilize laser processing to form laser processed regions that are isolated from the emitter. Laser processing may include laser-doping, laser-firing, laser-transfer, laser-transfer doping, laser contacting, and/or gas immersion laser doping.
US 9 761 739 B2 12 September 2017 High speed photosensitive devices and associated methods, SiOnyx LLC	High speed optoelectronic devices and associated methods are provided. In one aspect, for example, a high-speed optoelectronic device can include a silicon material having an incident light surface, a first doped region and a second doped region forming a semiconductive junction in the silicon material, and a textured region coupled to the silicon material and positioned to interact with electromagnetic radiation. The optoelectronic device has a response time of from about 1 picosecond to about 5 nanoseconds and a responsivity of greater than or equal to about 0.4 A W^{-1} for electromagnetic radiation having at least one wavelength from about 800 nm to about 1200 nm.

(*Continued*)

Table 9.3. (*Continued*)

Patent number/patent application publication number Date Patent/application title current assignee	Abstract
US 9 882 070 B2 30 January 2018 Photodetector structures and manufacturing the same, Aalto University Foundation	A photodetector structure comprises a semiconductor substrate extending substantially along a horizontal plane and having a bulk refractive index and a front surface defining the front side of the photodetector structure. The front surface comprises high aspect ratio nanostructures forming an optical conversion layer having an effective refractive index gradually changing towards the bulk refractive index to reduce reflection of light incident on the photodetector structure from the front side thereof. Further, the semiconductor substrate comprises an induced junction.
US 10 950 737 B2 16 March 2021 Semiconductor structures and manufacturing the same, Elfys Oy	A layered semiconductor structure with a width in a lateral direction, having an operating area covering part of the width of the semiconductor structure, comprises a semiconductor substrate with majority charge carriers of a first polarity; and a first dielectric layer with inducing net charge of the first polarity on the semiconductor substrate. An induced junction is induced in the semiconductor substrate by an electric field generated in the semiconductor substrate by the inducing net charge. The semiconductor structure is configured to confine the electric field generated in the semiconductor substrate in the operating area.
FI 127794 B 28 February 2019 Semiconductor structures and their preparation, Aalto Korkeakoulusaeaetioe	A layered semiconductor structure (100) with a width in a lateral direction, having an operating area (101) covering part of the width of the semiconductor structure, comprises a semiconductor substrate (102) with majority charge carriers of a first polarity; and a first dielectric layer (104) with inducing net charge of the first polarity on the semiconductor substrate. An induced junction (505) is induced in the semiconductor substrate by an electric field generated in the semiconductor substrate by

	the inducing net charge. The semiconductor structure is configured to confine the electric field generated in the semiconductor substrate in the operating area.
US 8 075 792 B1 13 December 2011 Nanoparticle-based etching of silicon surfaces, Alliance for Sustainable Energy, LLC	A method of texturing silicon surfaces such to reduce reflectivity of a silicon wafer for use in solar cells. The method includes filling a vessel with a volume of an etching solution so as to cover the silicon surface of a wafer or substrate. The etching solution is made up of a catalytic nanomaterial and an oxidant-etchant solution. The catalytic nanomaterial may include gold or silver nanoparticles or noble metal nanoparticles, each of which may be a colloidal solution. The oxidant-etchant solution includes an etching agent, such as hydrofluoric acid, and an oxidizing agent, such as hydrogen peroxide. Etching is performed for a period of time including agitating or stirring the etching solution. The etch time may be selected such that the etched silicon surface has a reflectivity of less than about 15% such as 1%–10% in a 350–1000 nm wavelength range.
US2012/0092390 A1* 19 April 2012 Low power image intensifier device comprising black silicon detector element, PFG IP LLC Irvine Sensors Corp	A detector module which may comprise a black silicon detector for detecting incident radiation and generating a visible output to a display output pixel element such as one or more OLED elements. One or more vertically interconnected unit cells, which may be in the form of a plurality of vertically stacked and bonded layers, is disclosed having a detector layer comprising one or more black silicon detector elements, an amplification layer and a display layer having a dedicated display output pixel element. The layers are vertically interconnected by means of electrically conductive area interconnects such as electrically conductive through-silicon vias.
W02011147122A1 5 August 2010 Black silicon solar cell and its preparation thereof https://patents.google.com/?assignee=...........&peid=6326d48c447d8:f2:1a07b277	Provided are a black silicon solar cell and its preparation method, which belong to the solar cell device manufacturing technology field. The black silicon solar cell includes a metal back electrode (1), the crystal silicon (2), a black silicon layer (3), a passivation

(*Continued*)

Table 9.3. (*Continued*)

Patent number/patent application publication number Date Patent/application title current assignee	Abstract
Institute of Microelectronics of CAS	layer (4) and a metal gate (5); wherein, the metal back electrode (1) is located on the back surface of the crystal silicon (2), the black silicon layer (3) is located on the crystal silicon (2), the passivation layer (4) is located on the black silicon layer (3), the metal gate (5) is located on the passivation layer (4). The preparation method includes: carrying out pretreatment of the silicon wafer; preparing the black silicon layer on the surface of the pretreated silicon wafer by using plasma submerged ion implantation technology; preparing an emitter on the black silicon layer, and carrying out passivation treatment on the emitter to form the passivation layer; respectively preparing the metal back electrode and the metal gate on the back surface of the single crystal silicon wafer and the passivation layer. The black silicon layer is utilized as an absorbing layer, so the average absorption rate of the cell is higher within the visible waveband.
CN115287606A 29 July 2022 A method for preparing a surface-enhanced Raman Spectroscopy substrate based on black silicon, XiAn Institute of Optics and Precision Mechanics of Chinese Academy of Sciences	The invention relates to a preparation method of an optical material, in particular to a preparation method of a surface enhanced Raman spectrum substrate based on black silicon, aiming at solving the problems that the prior art generally relates to more complicated material synthesis and processing technologies, and the realization of short time is difficult. The technical problem of large-scale preparation in time. The preparation method comprises the following steps: s1.1 at SF_6 In the atmosphere, irradiating by femtosecond laser to obtain black silicon; cutting the black silicon into black silicon slices by laser; s1.2, depositing noble metal films with different thicknesses on the surface of the black silicon

		slice in an evaporation coating mode to obtain a black silicon substrate; respectively immersing a plurality of black silicon substrates into 4-MBA standard ethanol solutions with different concentrations, and then taking out and drying; collecting the Raman spectrum of the surface of each black silicon substrate, and taking the thickness of the noble metal film on the surface of the black silicon substrate corresponding to the maximum value of the characteristic peak value of the 4-MBA molecule in the Raman spectrum as the optimal thickness of the noble metal film; *s*2, preparing the surface enhanced Raman spectrum substrate.
US 9 034 216 B2 19 May 2015 Wet-chemical systems and methods for producing black silicon substrates, Alliance for Sustainable Energy, LLC	A	wet-chemical method of producing a black silicon substrate. The method comprising soaking single crystalline silicon wafers in a predetermined volume of a diluted inorganic compound solution. The substrate is combined with an etchant solution that forms a uniform noble metal nanoparticle induced Black Etch of the silicon wafer, resulting in a nanoparticle that is kinetically stabilized. The method comprising combining with an etchant solution having equal volumes acetonitrile/acetic acid: hydrofluoric acid:hydrogen peroxide.
US 8 828 765 B2 9 September 2014 Forming high efficiency silicon solar cells using density-graded anti-reflection surfaces, Alliance for Sustainable Energy, LLC	A	method is provided for processing a graded-density AR silicon surface to provide effective surface passivation. The method includes positioning a substrate or wafer with a silicon surface in a reaction or processing chamber. The silicon surface has been processed to be an AR surface with a density gradient or region of black silicon. The method continues with heating the chamber to a high temperature for both doping and surface passivation. The method includes forming, with a dopant-containing precursor in contact with the silicon surface of the substrate, an emitter junction proximate to the silicon surface by doping the substrate. The method further includes, while the chamber is maintained at the high or raised temperature, forming a passivation layer on the graded-density silicon anti-reflection surface.

(*Continued*)

Table 9.3. (*Continued*)

Patent number/patent application publication number Date Patent/application title current assignee	Abstract
JP6868908B2 12 May 2021 Structured silicon-based thermal emitter, Si Ware Systems, Si Ware Systems Inc.	This PCT International Patent Application claims the priority of the following patent applications, which are incorporated herein by reference in their entirety and in fact form part of this PCT International Patent Application. doing. 1. US Provisional Application No. 62/191, 254 'Structured Silicon-Based Thermal Emitter' (agent registration number SIWA-1023PROV) filed on July 10, 2015, pending application. 2. US Patent Application No. 15/203,773 'Structured Silicon-Based Thermal Emitter' (agent registration number SIWA-1023) filed July 6, 2016, pending application. Aspects of the invention generally relate to optical sources, especially to silicon-based thermal emitters.
US 8 790 953 B2 29 July 2014 Method for texturing silicon surface to create black silicon for photovoltaic applications, Blacksilicon Ltd	The surface of silicon is textured to create black silicon on a nano-micro scale by electrochemical reduction of a silica layer on silicon in molten salts. The silica layer can be a coating, or a layer caused by the oxidation of the silicon.
W02006086014A3 23 September 2005 Method for manufacturing of silicon-based detektors having laser-microstructured sulfur-doped surface layers, Harvard College, Eric Mazur, James Edward Carey III	The present invention provides methods of fabricating a radiation-absorbing semiconductor wafer by irradiating at least one surface location of a silicon substrate, e.g., an n-doped crystalline silicon, by a plurality of temporally short laser pulses, e.g., femtosecond pulses, while exposing that location to a substance, e.g., SF6, having an electron-donating constituent so as to generate a substantially disordered surface layer (i.e., a microstructure layer) that incorporates a concentration of that electron-donating constituent e.g., sulfur. The

substrate is also annealed at an elevated temperature and for a duration selected to enhance the charge carrier density in the surface layer. For example, the substrate can be annealed at a temperature in a range of about 500 K to about 1000 K.

* Abandoned.
** Expired.

9.5 Summary

Since the discovery of BSi by Eric Mazur and his team at Harvard University in the 1990s, it continues to see significant growth. Some of the major applications of BSi are as follows:

- Photovoltaics, due to the enhanced light absorption and hence the potential for increased solar cell efficiency.
- Semiconductor devices such as photodetectors, imagers, and other optoelectronic devices.
- Medical and biotech applications including biosensors, drug delivery systems, and diagnostic devices.
- Consumer electronics applications such as advanced sensors and solar cells in smart phones and wearables.
- Optics and sensing applications—examples include the design and development of instrumentation and advanced imaging systems for a variety of applications including astronomy and defense, light detection and ranging (LIDAR) in autonomous vehicles, and terahertz emitters.

References

[1] Jansen H, de Boer M, Legtenberg R and Elwenspoek M 1995 The black silicon method: a universal method for determining the parameter setting of a fluorine-based reactive ion etcher in deep silicon trench etching with profile control *J. Micromech. Mecroeng.* **5** 115–20

[2] Tsing-Hua H, Finlay R J, Wu C, Deliwala S and Mazur E 1998 Microstructuring of silicon with femtosecond laser pulses *Appl. Phys. Lett.* **7** 1673–5

[3] Autm Better World Project Black silicon revolutionizes how light is captured and stored *Harvard University* https://autm.net/about-tech-transfer/better-world-project/bwp-stories/black-silicon-revolutionizes-how-light-is-captured (Accessed: 5 April 2025)

[4] Xiaogang L, Coxon P R, Marius P, Bram H, Cole J M and Fray D J 2014 Black silicon: fabrication methods, properties and solar energy applications *Energy Environ. Sci.* **7** 3223–63

[5] Fan Z *et al* 2021 Recent progress of black silicon: from fabrications to applications *Nanomaterials* **11** 41

[6] Zhao Z *et al* 2023 Black silicon for near-infrared and ultraviolet photodetection: a review *APL Mater.* **11** 021107

[7] Crawford M 2019 Black silicon: 23 years later *SPIE News* https://spie.org/news/spie-professional-magazine-archive/2019-april/black-silicon (Accessed: 5 April 2025)

[8] Harvard School of Engineering 2008 Novel material black silicon is licensed by Harvard *Harvard School of Engineering News* https://seas.harvard.edu/news/2008/10/novel-material-black-silicon-licensed-harvard (Accessed: 5 April 2025)

[9] SIONYX https://sionyx.com/ (Accessed: 5 April 2025)

[10] Yee K and White V Black silicon blacker than black *NASA Jet Propulsion Laboratory, California Institute of Technology* https://microdevices.jpl.nasa.gov/capabilities/advanced-microfabrication-technologies/black-silicon/ (Accessed: 6 April 2025)

[11] Natcore scientists build world's first black silicon solar cell using scalable production process *PV Magazine* 25 October 2012 https://pv-magazine.com/press-releases/natcore-scientists-build-worlds-first-black-silicon-solar-cell-using-scalable-production-process_10008981/ (Accessed: 6 April 2025)

[12] Coveme sign joint development agreement *Natcore Technology* 4 April 2017 https://newswire.ca/news-releases/natcore-technology-coveme-sign-joint-development-agreement-618160303.html (Accessed: 6 April 2025)

[13] Andorka F 2017 Natcore's foil cell creeps closer to commercialization *PV Magazine* 5 April https://pv-magazine-usa.com/2017/04/05/natcores-foil-cell-creeps-closer-to-commercialization/ (Accessed: 6 April 2025)

[14] Natcore technology develops process to manufacture solar cells without silver *Power Technology* 9 April 2015 https://www.power-technology.com/news/newsnatcore-technology-develops-process-to-manufacture-solar-cells-without-silver-4652012/ (Accessed: 6 April 2025)

[15] Carlson D E and Levy D H 2019 Systems and methods for forming foil contact rear emitter solar cells with carrier selective contacts *US Patent* 10 453 976 B2

[16] Levy D H and Carlson D E 2017 Laser processed back contact heterojunction solar cells *US Patent* 9 837 561 B2

[17] Oh J, Yuan H-C and Branz H M 2012 An 18.2%-efficient black-silicon solar cell achieved through control of carrier recombination in nanostructures *Nat. Nanotechnol.* **7** 743–8

[18] NREL 2010 NREL's black silicon increases solar cell efficiency by reducing reflected light *Report* NREL/FS-6A4-48772 National Renewable Energy Laboratory https://nrel.gov/docs/fy11osti/48772.pdf (Accessed: 7 April 2025)

[19] Oh J, Branz H M and Yuan H-C 2022 Efficient black silicon photovoltaic devices with enhanced blue response *US Patent* 11 251 318 B2

[20] Schmelz D *et al* 2023 Investigations on black silicon nanostructures fabricated by reactive ion etching on highly curved surfaces *Mater. Sci. Semicond. Process.* **165** 107646

[21] Photodiodes *Thor Labs* https://www.thorlabs.com/newgrouppage9.cfm?objectgroup_id=285 (Accessed: 8 April 2025)

[22] Savin H *et al* 2015 Black silicon solar cells with interdigitated back-contacts achieve 22.1% efficiency *Nat. Nanotechnol.* **10** 624–8

[23] Garin M *et al* 2020 Black-silicon ultraviolet photodiodes achieve external quantum efficiency above 130% *Phys. Rev. Lett.* **125** 117702

[24] Forbes L 2025 The history of 'black silicon'in photodetector patents *Forbes4* http://www.forbes4.com/THE HISTORY OF BLACK SILICON IN PHOTODETECTOR PATENTS.pdf (Accessed: 8 April 2025)

Chapter 10

Summary—R&D, products, and opportunities

Black silicon (BSi), a surface modification of silicon, exhibits low reflectance and correspondingly high absorptance in the visible and infrared range of wavelengths, making it more viable than crystalline silicon for applications in silicon photonics. This property, coupled with its large surface area due to its morphology, has led to numerous studies that have been aimed at investigating a variety of applications. Most importantly, its utility in image sensors, gas sensors, and solar cells has been examined extensively in the literature. On a fundamental level, BSi offers a combination of the best of both worlds—the high absorption of black materials and the semiconducting properties of silicon. Furthermore, the use of BSi in photonics and solar cells circumvents the need for antireflection coatings since BSi plays the dual role of a photon absorbing layer while at the same time performing the role of an antireflection coating.

10.1 Introduction

It has been thirty years since Jensen *et al* reported 'The black silicon method: a universal method for determining the parameter setting of a fluorine-based reactive ion etcher in deep silicon trench etching with profile control' [1]. An illustration of the synergistic mechanism of the chemistry of the plasma, $SF_6/O_2/CHF_3$, deployed by Jensen *et al* for etching silicon, with a controllable etch profile, is presented in figure 10.1 [2].

The chemical etching of silicon occurs by the formation of volatile SiF_4 by the F^- radicals that are produced by the SF_6 plasma. The flow rate of these gases, SF_6, O_2, and CHF_3, controls the etch rate of silicon as well as the etch profiles [1, 2]. The first results of the successful demonstration of black silicon (BSi) [3] as a new light absorber for applications in photovoltaics [4, 5] were reported by Eric Mazur and his research team at Harvard University. Subsequently, the applications of BSi were extended to visible and near-infrared optoelectronic devices [6]. These devices are manufactured by SIOnyx, a silicon-based photonics company cofounded by Eric

doi:10.1088/978-0-7503-3247-7ch10

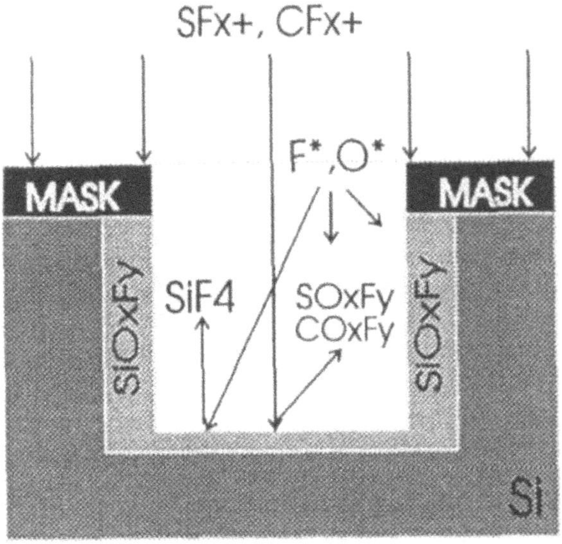

Figure 10.1. The $SF_6/O_2/CHF_3$ chemistry [1]. (Reproduced with permission from [2]. Copyright 1995 The Electrochemical Society.)

Mazur and James Carey [7]. Recently, SIOnyx acquired Amigen, a thermal imaging systems provider [8].

An example of the evolution of the morphology of BSi is presented in figure 10.2 [9]. The experimental procedure for the formation of the structures, shown in figure 10.2, is described in detail in [9]. These images, acquired by utilizing scanning electron microscopy (SEM), show varying pitches of (a) 1400 nm, (b) 800 nm, (c) 500 nm, and (d) 300 nm, with the samples tilted at 70° angle from overhead [9].

10.2 Examples of R&D, products, and opportunities

The applications of BSi are primarily determined by its properties which in turn are highly sensitive to processes. In chapter 4, the details of a variety of processes for the manufacture of BSi are discussed. BSi provides ample opportunities for R&D and process optimization. In chapter 6, the numerous applications of BSi are discussed briefly. In general, the applications of BSi, in the form of materials and devices, include the following:

- Antibacterial surfaces.
- Antireflection coatings.
- Catalysis.
- Drug delivery systems.
- Electrooptics.
- Field emission.
- Gas sensors.
- Gettering of metallic impurities.
- Hydrogen generation.

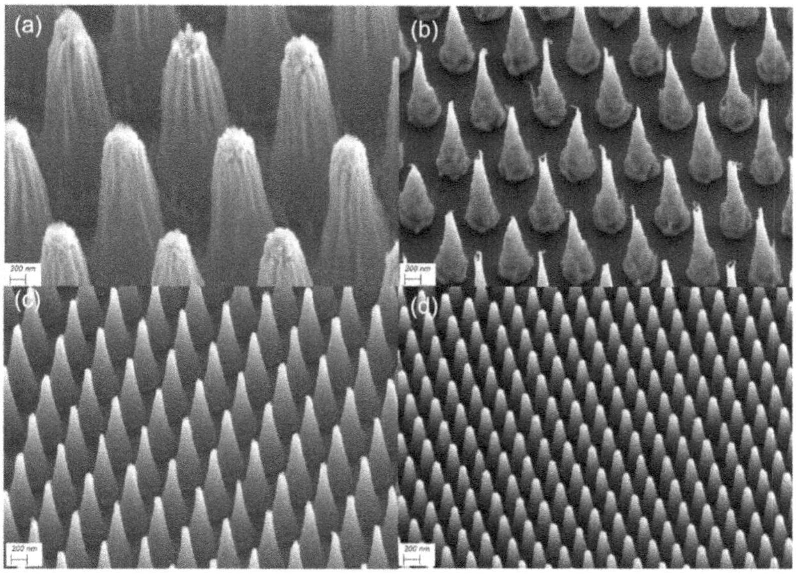

Figure 10.2. SEM images of fabricated BSi substrates. Pitches of (a) 1400, (b) 800, (c) 500, and (d) 300 nm. Scale bar: 200 nm. (Reproduced from [9]. Copyright 2024 the Authors, published by IOP Publishing, CC BY 4.0.)

- Infrared/photo detectors/imagers.
- Microfluidics.
- Optical devices.
- Passivation.
- Photodetectors.
- Radiative cooling.
- Sensors.
- Solar cells.
- Supercapacitors.
- Thermal management.
- Thermophotovoltaics.
- THz-range radiation emitters.

As can be seen from this exhaustive list, the applications of BSi are numerous. However, it must be noted that, at this time, only a few of the above-mentioned applications have been converted into commercial products. SiOnyx continues to be the world leader in the use of BSi for the manufacture of imagers with a focus on its proprietary ultra-low-light CMOS image sensors and high-performance night-vision camera systems [10]. This capability differentiates SiOnyx from its competitors, such as Forward Looking Infra-Red (FLIR) Teledyne [11], Photonis [12], and CONTROP Precision Technologies [13].

10.2.1 Advantages and disadvantages of BSi

The surface roughness of BSi, at the nanoscale, offers several advantages, particularly in its ability to enhance light absorption over a broad range of wavelengths. This has direct implications for improved solar cell performance and efficiency. However, it also has disadvantages such as higher manufacturing costs and increased surface recombination. BSi is an excellent candidate for assessment of the return on investment (ROI) and cost–benefit analysis [14–17]. It is also a good material from the perspective of understanding its process–property–performance correlations.

The advantages of using BSi include the following.

Enhanced light absorption [18, 19]:
The rough surface of BSi increases its surface area, allowing it to absorb more light, which is of interest to device applications such as solar cells.

Large and chemically active surface area [20]:
The rough surface of BSi provides a larger and more reactive surface area due to its high aspect ratio, which can be useful for a variety of applications, including catalysis [20] and sensing [21, 22].

Photodetection enhancement [19, 23, 24]:
BSi can improve the sensitivity of photodetectors, especially in the near-infrared and ultraviolet ranges, due to its increased surface area and ability to trap photons and lead to the increased generation of electron–hole pairs.

Potential for higher efficiency in solar cells [25, 26]:
BSi solar cells can achieve higher efficiencies than conventional silicon cells due to the enhanced light absorption and reduced reflection.

Superhydrophobicity [27, 28]:
BSi surfaces can exhibit superhydrophobicity, making them water-repellent, which can be beneficial in certain applications.

The disadvantages of the use of BSi comprise the following [29, 30].

Potential metal contamination:
The metal-assisted wet chemical etching (MACE) process for the fabrication of BSi has been shown to result in metal contamination of silicon [15].

Minority carrier lifetime:
The minority carrier lifetime in BSi is affected by several experimental parameters including the fabrication method, etching time, and the presence of defects that may be potentially introduced during processing.

10.2.2 Case studies

For all practical purposes, BSi may be considered as an excellent example of 'material = device'. While the origin of BSi dates back to the 1980s [31], the transition from fundamental research to applications of BSi and subsequent prototyping, product development, and manufacturing was generally slow. 'The black silicon method' described by Jensen *et al* [1, 2], followed by the pioneering work of Mazur and his team [3–6], laid the foundation for the growth of research in

BSi. One of the major advantages of BSi is its ability to provide an all-silicon solution from the fundamental goal of facile integration of silicon photonics with silicon microelectronics. Indeed, the products from SiOnyx are excellent examples of this all-silicon approach.

Some of the applications of BSi are described in the following sections.

10.2.2.1 New functionalities in microsystems

Micro-electromechanical systems (MEMS) technology facilitates the fabrication of very small devices and is generally considered to be a cost-effective approach for the fabrication of devices, circuits, and systems. Examples of these include 'XYZ on a chip', 'lab on a chip', 'systems on a chip', microfluidics etc. MEMS based components are generally of dimensions in between 1 and 100 μm. BSi is found to be a useful material for mechanical processes such as bonding, as a carrier for catalytic materials, etc. The ability to fix micro parts helps in novel methods for the assembly of hybrid micromechanical systems, sensors, and actuator arrays [32–35].

In addition to the applications in low-temperature bonding technologies, the silicon needles that are formed in BSi are of interest to micro-reaction technology. BSi can be coated with various functional layers and is therefore an attractive candidate for use in chemical reaction systems. Figure 10.3 presents an example of a microreactor system with a platinum catalyst on BSi [32].

10.2.2.2 Solar cells

Solar energy is the leading source of renewable energy in the world today. Solar photovoltaics (PV) has been the major contributor to the growth of renewable energy, accounting for 70% of the increase in renewable electricity generation in 2023. Among the fabrication technologies of solar cells, the manufacture of silicon solar cells is the most mature and makes up ~95% of the global solar cell market. The installed $/watt and the reliability of silicon solar panel performance, combined with the abundance of silicon in nature, are some of the factors that

Figure 10.3. Micro-reaction system with platinum catalyst on BSi (chip size: 41 mm × 27 mm) [32]. (Reproduced with permission from [32]. Copyright 2006 IOP Publishing.)

have contributed to the continued dominance of silicon in the solar PV industry. However, the indirect energy gap of silicon (~1.12 eV at room temperature) and its reflectance of ~30%–35% in the visible range of wavelengths are the limiting factors for application in solar cells. The nanostructured surface of BSi behaves as an excellent antireflection coating to enhance light harvesting in photovoltaics [36–40]. Thus the use of BSi in the manufacture of solar cells circumvents the need for any antireflection coating.

The crystalline silicon wafers that are used to manufacture today's silicon solar cells are processed to form a textured surface [41] to be able to trap incident photons. The wafers are then coated with an antireflective layer, usually silicon nitride, using high-vacuum processes [42]. This antireflection coating layer enhances the performance of the underlying silicon solar cells by decreasing the reflectance, increasing thereby the photon-induced electron–hole pairs in the cell. However, the antireflection coating also adds to the costs of the process.

Since its inception, BSi has been a subject of significant interest in various areas of research and applications, including photovoltaics, for its ability to reduce the surface reflectance even below 1%. The advancement in the performance of BSi solar cells was reported in 2013 with the solar cell efficiency reported to be 18.7% by a research group at Aalto University, Finland, and Fraunhofer ISE, Germany [43]. In 2015, researchers at Aalto University, Finland, set a new record for developing the most efficient solar cells. The nanostructured BSi solar cells exhibited an efficiency of 22.1% [44]. The increase in efficiency was achieved by incorporating a conformal alumina film, via atomic layer deposition (ALD), to an interdigitated back contact (IBC) cell's front surface (shown in figure 10.4), to reduce surface recombination. The conformal alumina film reduced the surface recombination in BSi solar cells by providing excellent chemical and electrical passivation [44, 45]. The previous efficiency record of 18.2% for BSi solar cells was held by the US Department of Energy's National Renewable Energy Laboratory (NREL) using thermal oxide as a passivating layer [46]. The present efficiency record for BSi solar cells is 22.1%, held by Aalto University in Finland [44, 47].

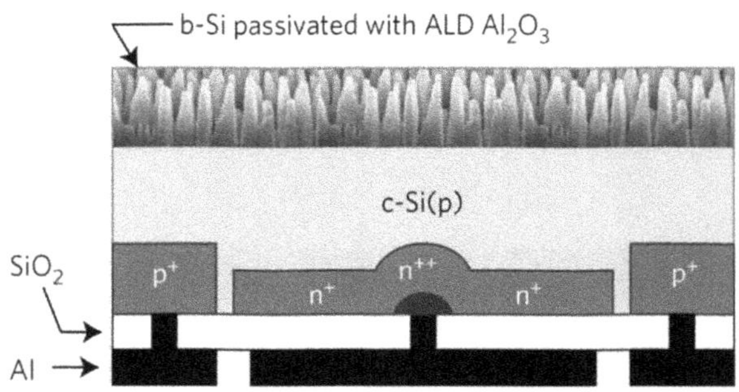

Figure 10.4. Structure of the interdigitated BSi solar cell. A thin Al_2O_3 layer is deposited on the nanostructured front surface. (Reproduced with permission from [44]. Copyright 2015 Springer Nature.)

10.2.2.3 Image sensors

Since the successful demonstration of the processing of texturing silicon surfaces utilizing femtosecond laser pulses [5], and the subsequent formation of BSi, Eric Mazur and James Carey of Harvard University founded the startup SiOnyx in 2006. The company, based in Beverly, MA, uses its proprietary BSi laser process to improve the infrared sensitivity of silicon-based photodetectors and image sensors in low-light and night-vision applications, that are suitable for a wide range of industries [7]. SiOnyx XQE™ CMOS image sensors provide superior night vision, biometrics, eye-tracking, and natural human interfaces through their proprietary BSi semiconductor technology [48]. XQE image sensors have been reported to deliver unprecedented performance advantages in infrared imaging, including sensitivity enhancements as high as ten times that of today's sensor solutions. XQE enhanced IR sensitivity takes advantage of the naturally occurring IR 'nightglow' to enable imaging under extreme (< 0.001 lux) conditions. XQE sensors provide high quality daytime color as well as nighttime imaging capabilities that offer new levels of performance and threat detection. Systems that depend upon IR light to perform will be redefined by XQE sensor technology—from longer range covert IR illumination, to industrial IR sensing and range finding. The enhanced IR sensitivity of XQE sensors enables higher performance and lower power IR based solutions at wavelengths > 900 nm.

10.2.2.4 Surface-enhanced Raman spectroscopy

Raman spectroscopy is a powerful technique to detect and analyse chemical compounds. It can be used in biomedical detection, toxin analysis, and environmental monitoring. Poor signal intensity from Raman scattering is the main issue in the detection of analytes with low concentration. In order to address this issue, surface enhanced Raman spectroscopy (SERS) has been proposed as an excellent characterization technique for material analyses. The Raman signal enhancement factor in SERS can be as high as 10^6–10^8. BSi is ideal for SERS due to its distinct properties such as low reflectance, large surface contact area, and sharp tips [49, 50].

The major advantage of BSi in SERS is its broadband and omnidirectional enhancement due to its irregular corrugated surface structure. For a normal silicon nanostructure, i.e. usually a periodic structure, the coupling is highly wavelength and angle selective. Therefore, the normal silicon nanostructure is usually iridescent. But BSi looks black from all directions. The implication is that BSi can efficiently absorb light in very broad bandwidths from wide angles. Even with metal deposited on BSi (which is supposed to give rise to a mirror surface), it still couples light from a broad bandwidth and wide angles. The absorption or coupling in BSi can be attributed to two factors. One is the gradient effective refractive index of the sharp nanocone layer. The other is the diffraction of the irregular sub-wavelength nanocone array. Due to randomness in the structure, there is the possibility of overwhelming interference (constructive) of surface plasmons at some location which will give rise to a very high electromagnetic field ('hot spot'). The irregular corrugated nanostructure also provides an additional surface plasmon coupled scattering path for the photons. All these factors contribute to the advantages of

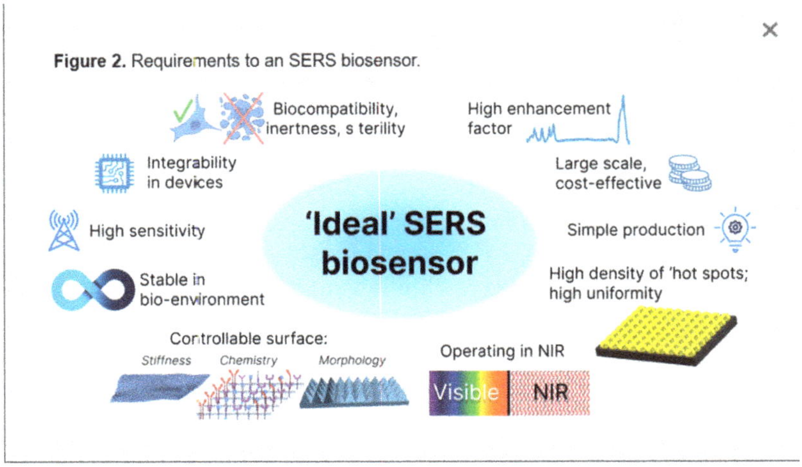

Figure 10.5. Requirements for an SERS biosensor. The requirements are described in detail in [52]. (Reproduced from [52]. Copyright 2024 MDPI, CC BY 4.0.)

using BSi for applications in SERS [51]. See figure 10.5 for the requirements for an SERS biosensor.

10.2.2.5 Biosensors

In the natural environment, cells interact with nanostructures. Based on this, nanopatterned implant surfaces provide better adhesion and biocompatibility. BSi comprises numerous nanostructures [53]. Several investigations are in progress in this area of research. BSi can also be used for sensing and fingerprinting of molecular and microbial contamination [54]. This same property can be exploited in chemical analysis and biosensing applications [55, 56]. The enhanced surface area and excellent light absorption of BSi facilitates in the detection of small amounts of analytes. Specific biological molecules can be detected and targeted by functionalizing BSi with specific molecules or antibodies. Overall, BSi fabrication methods are completely compatible with existing silicon technologies. This makes them cost-effective and scalable. The nanostructure of BSi permits the fabrication of miniaturized biosensors.

10.2.2.6 Gas sensors

Gas detection is important in various areas of application, such as industrial emission control, household security, vehicle emission control, etc. Various semiconductor metal oxides [57–60] have been used with significant impact in gas detection. Due to the high surface area and volume ratio, nanostructured materials can be used for gas detection. This is also helpful in adsorption of gases. They are safer to operate and consume less power. BSi has all these advantages and can be used as a gas sensor.

Figure 10.6 presents an illustration of the adsorption and reaction mechanism of the sensing process on the surface of a BSi-based gas sensor [61].

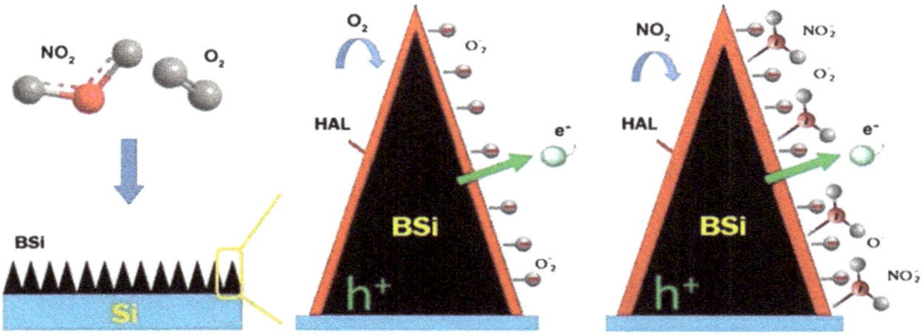

Figure 10.6. Adsorption and reaction mechanism of the sensing process on the BSi-based sensor surface. (Reproduced with permission from [61]. Copyright 2023 Wiley.)

10.2.2.7 Anti-bacterial surfaces

Researchers at the Swinburne University of Technology in Australia have discovered that BSi has the capability of killing bacteria [62, 63]. It has been reported that this finding will pave the way for new nanostructured antibacterial materials that can be used in various medical applications such as the surfaces of medical implants [64, 65].

BSi, with its needle-shaped nanostructures, which is used primarily for sensor applications, also serves as a potent antibacterial agent, killing some 450 000 cells per minute in just one square centimeter [64]. Studies have shown that certain nanostructures can indeed kill bacteria based on texture alone [66]. It had been previously demonstrated that cicada (*Psaltoda claripennis*) wings, which are covered in dense nanopillar structures, were highly lethal to the opportunistic human pathogen *Pseudomonas aeruginosa* and provided evidence to suggest that the wings' biochemical properties were not responsible for such activity. It was demonstrated that the bactericidal nature of the wing is due to the mechanical rapture of bacterial cells [67–69]. Recognizing that dragonflies (*Diplacodes bipunctata*) also had similar nanopillars on their wings, and that BSi is known for a similar nano-texture, researchers have explored the bactericidal potential of the two surfaces. Both caused significant deformation of the cell walls of *P. aeruginosa*, *Staphylococcus aureus*, and *Bacillus subtilis*, and even killed *B. subtilis* spores—at rates more than sufficient to stave off the bacteria [64]. They provided potent bactericidal activity against not only gram-negative bacteria but also against the more rigid and lysis-resistant gram-positive bacteria and their spores. Researchers continue to explore a range of other nanomaterials/nanocoatings whose surfaces maybe suitable for the fabrication of similar structural nano-patterns to create surfaces free from bacteria [69].

10.2.2.8 Drug delivery

Modern day developments in biotechnology have facilitated many sophisticated and powerful drugs. Many of the drugs are more effective when delivered using transdermal drug delivery [70]. Skin has poor permeability, necessitating micro-needles that are not only long enough to reach the deeply placed nerves in the human body, but also cause less pain in this whole process. BSi, prepared by reactive ion

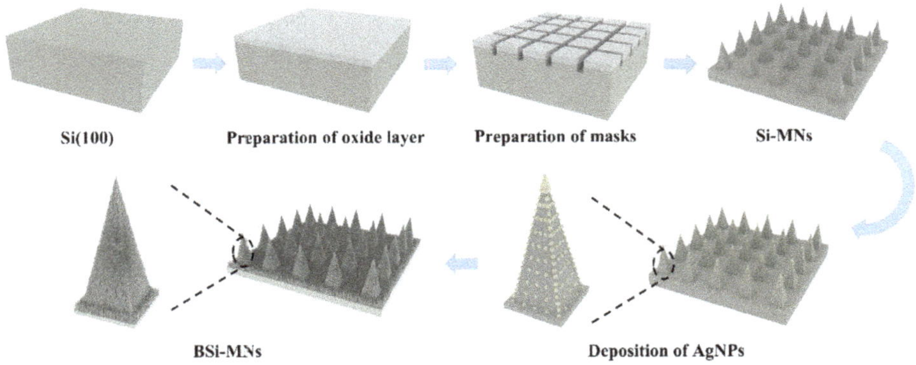

Figure 10.7. Schematic diagram of the formation of a BSi-MN array. (Reproduced from [73]. Copyright 2023 MDPI, CC BY 4.0.)

etching (RIE), has been used for transdermal drug delivery successfully and with notably less pain [71, 72]. Generally, the BSi microneedles have two important features: first, they have extremely sharp tips (a radius of curvature <1 μm) which facilitate easy piercing into the skin. Second, they can be made approximately 150 μm long. Because the skin surface is not flat due to dermatoglyphics (i.e. wrinkles) and hair, the full length of these microneedles will not penetrate the skin. The part that does not penetrate should insert deep enough to cross the stratum corneum barrier but not so deep to stimulate nerves found in the dermis. The microfabrication technique, used in BSi fabrication, can easily be modified to make longer or shorter needles if needed. These microneedles do not break when inserted into the skin and are capable of increasing skin permeability by up to more than four orders of magnitude. The RIE technique used to fabricate these microneedles is relatively simple and is therefore amenable to inexpensive mass production for future applications. This approach has the potential to significantly increase the number and types of drugs which can be delivered across the skin [71, 72].

Cheng *et al* [73] have reported a study on silicon microneedles (MNs) that were prepared by utilizing laser patterning and alkaline etching. Subsequently, BSi-MNs were formed by a process of silver-catalyzed chemical etching. An illustration of their process is shown in figure 10.7 [73].

10.2.2.9 Terahertz emitter

The region between microwave and infrared has become a more interesting area of research due to the latest advent in ultrashort pulse lasers. Terahertz radiation and detection has applications in a wide range of areas such as security scanning, quality control and space research. BSi can emit terahertz (THz) radiation when excited by short optical pulses. BSi has been used as a terahertz emitter in several studies in literature [74, 75].

Upon irradiation from ultrashort pulses from a Ti:sapphire laser, terahertz emission is observed from the BSi emitter. An illustration of the experimental setup is presented in figure 10.8 [75]. It may be noted that a polished silicon wafer does not show any detectable terahertz emission.

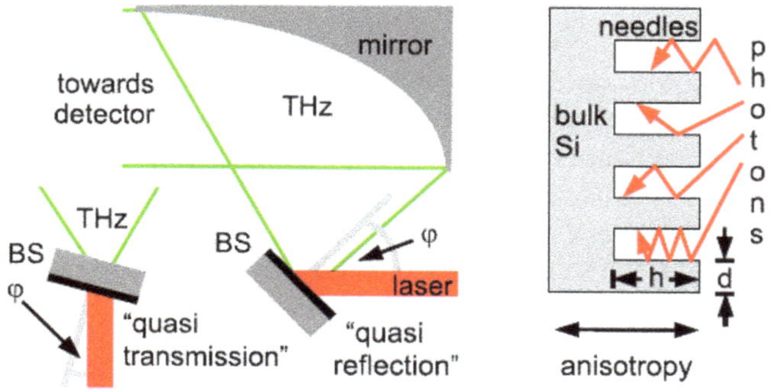

Figure 10.8. Experimental set-up. Left: BSi emitter in quasi-transmission. Middle: Quasi-reflection geometry. Right: Possible photon paths. (Reproduced with permission from [75]. Copyright 2008 AIP.)

10.2.2.10 Other applications

BSi exhibits a superhydrophobic property [27, 76]. This is particularly useful in developing self-cleaning surfaces, particularly for applications in the automobile, aviation, electronic, packaging, marine, and textile industries. Other applications of this characteristic of BSi include the development of anti-biofouling in medical devices as well as in solar energy harvesters [27]. BSi has been known to be a highly efficient photo-thermal converter for snow/ice melting [77] as well as efficient in evaporating water [78]. The property of snow/ice melting has been shown to be useful for applications in agriculture. With an emissivity of $\sim$1 (0.988 or higher in the wavelength range of 3–25 μm), almost equal to that of a black body, BSi is extremely useful in thermal management, solar thermal conversion, and passive radiative cooling, as well as in the cooling of high-power electronics. Due to its excellent absorption in the infrared range of wavelengths, BSi enhances the performance of thermophotovoltaic cells in their ability to convert infrared radiation into electricity [79, 80]. The properties of BSi have been taken advantage of in various characterization techniques such as machine vision systems [81] and scanning probe microscopy [82]. The optical properties of BSi have been utilized by NASA to design highly absorptive pupil masks [83].

10.3 Conclusions

BSi continues to attract the attention of engineers and scientists to address a variety of its applications. These applications include energy, healthcare, semiconductors, thermal management, etc. The primary driver for this continued interest can be attributed to the morphology of BSi coupled with its optical properties. The opportunities for product development, based on BSi are significant. BSi-based photonics, coupled with silicon microelectronics, is expected to penetrate new areas of the industry, particularly in aerospace, defense, and energy.

References

[1] Jansen H, de Boer M, Legtenberg R and Elwenspoek M 1995 The black silicon method: a universal method for determining the parameter setting of a fluorine-based reactive ion etcher in deep silicon trench etching with profile control *J. Micromech. Microeng.* **5** 115–20

[2] Legtenberg R, Jansen H, de Boer M and Elwenspoek M 1995 Anisotropic RIE of Si using $SF_6/0_2/CHF_3$ gas mixtures *J. Electrochem. Soc.* **142** 2020–8

[3] Her T-H, Finlay R J, Wu C, Deliwala S and Mazur E 1998 Microstructuring of silicon with femtosecond laser pulses *Appl. Phys. Lett.* **73** 1673–5

[4] Wu C, Her T-H and Mazur E 1998 Black silicon: a new light absorber for photovoltaic applications *Conf. on Lasers and Electro-Optics* ed D Scifres and A Weiner (Optica Publishing Group) paper CPD18

[5] Wu C, Her T and Mazur E 1999 Black silicon: new light absorber *APS Centennial Meeting (Atlanta, GA)* https://mazur.harvard.edu/sites/projects.iq.harvard.edu/files/mazur/files/talk_265.pdf

[6] Mazur E and Carey J E 2016 Silicon-based visible and near-infrared optoelectronic devices *US Patent* 9 276 143 B2

[7] SIONYX https://sionyx.com/ (Accessed: 21 April 2025)

[8] Bowman C 2023 SIONYX acquires Amigen *SIONYX* www.sionyx.com/blogs/news/sionyx--acquires-amigen (Accessed: 21 April 2025)

[9] Kayes M I, Zarei M, Feng F and Leu P W 2024 Black silicon spacing effect on bactericidal efficacy against gram-positive bacteria *Nanotechnology* **35** 025102

[10] Our mission *SIONYX* https://sionyx.com/pages/about-sionyx (Accessed: 23 April 2025)

[11] Teledyne FLIR https://flir.com/ (Accessed: 23 April 2025)

[12] Photonis Defense https://photonisdefense.com/ (Accessed: 23 April 2025)

[13] Controp https://controp.com/ (Accessed: 23 April 2025)

[14] Edwin O 2025 This dark material is the end of solar panels: it's about to change the world *EcoNews* https://ecoticias.com/en/black-silicon-is-the-end-solar-panels/10308/ (Accessed: 26 April 2025)

[15] Soueiti J *et al* 2023 A review of cost-effective black silicon fabrication techniques and applications *Nanoscale* **15** 4738–61

[16] Modanese C *et al* 2018 Economic advantages of dry-etched black silicon in passivated emitter rear cell (PERC) photovoltaic manufacturing *Energies* **11** 2337

[17] Golubewa L *et al* 2023 Black silicon: breaking through the everlasting cost vs effectivity trade-off for SERS substrates *Materials* **16** 1948

[18] Marthi S R, Sekhri S and Ravindra N M 2015 Optical properties of black silicon: an analysis *JOM* **67** 2154–9

[19] Pasanen T 2025 Black silicon offers enhanced responsivity to overcome near-infrared photodetection challenges *Photonics Spectra* https://photonics.com/Articles/Black_Silicon_Offers_Enhanced_Responsivity_to/p5/a70766 (Accessed: 23 April 2025)

[20] Ali M *et al* 2016 Nanostructured photoelectrochemical solar cell for nitrogen reduction using plasmon-enhanced black silicon *Nat. Commun.* **7** 11335

[21] Lv J *et al* 2018 Review application of nanostructured black silicon *Nanoscale Res. Lett.* **13** 110

[22] Dong B *et al* 2024 Light and gas dual-function detection and mutual enhancement based on hyperdoped black silicon *Opt. Express* **32** 13384–95

[23] Wang X *et al* 2024 Femtosecond laser-induced black silicon: a dual application in photodetection and surface-enhanced Raman scattering *Opt. Express* **32** 36362–80

[24] Zhao Z *et al* 2023 Black silicon for near-infrared and ultraviolet photodetection: a review *APL Mater.* **11** 0211107

[25] Savin H *et al* 2015 Black silicon solar cells with interdigitated back-contacts achieve 22.1% efficiency *Nat. Nanotechnol.* **10** 624–8

[26] Xu J *et al* 2021 High-efficiency black silicon tunnel oxide passivating contact solar cells through modifying the nano-texture on micron-pyramid surface *Sol. Energy Mater. Sol. Cells* **233** 111409

[27] Meng L *et al* 2025 Enhanced superhydrophobic robustness of black silicon employing nanojungle structures *Nanoscale* **17** 2871–7

[28] Ayvazyan G *et al* 2024 Wetting properties of black silicon layers fabricated by different techniques *Phys. Status Solidi—Rapid Res. Lett.* **18** 2400072

[29] Miakonkikh A and Kuzmenko V 2024 Formation of black silicon in a process of plasma etching with passivation in a SF_6/O_2 gas mixture *Nanomaterials* **14** 945

[30] Davidsen R S *et al* 2018 Black silicon realized by reactive ion etching (ICP) without platen power *IEEE 7th World Conf. on Photovoltaic Energy Conversion (WCPEC) (A Joint Conf. of 45th IEEE PVSC, 28th PVSEC and 34th EU PVSEC)* (Piscataway, NJ: IEEE)

[31] Flory F R, Le Rouzo J, Escoubas L and Berginc G 2017 Nanophotonics: from fundamental research to applications *Micro- and Nanophotonic Technologies* (New York: Wiley) ch 1

[32] Stubenrauch M *et al* 2006 Black silicon—new functionalities in microsystems *J. Micromech. Microeng.* **16** S82–7

[33] Jansen H *et al* 1996 The black silicon method VI: high aspect ratio trench etching for MEMS applications *Proc. Ninth Int. Workshop on Micro Electromechanical Systems* (Piscataway, NJ: IEEE) pp 250–7

[34] Barillaroa G *et al* 2002 Electrochemical etching in HF solution for silicon micromachining *Sensors Actuators* A **102** 195–201

[35] Marthi S R 2017 Black silicon—optical properties and applications *PhD Dissertation* (New Jersey Institute of Technology)

[36] Dhoska K, Spaho E and Sinani U 2024 Fabrication of black silicon antireflection coatings to enhance light harvesting in photovoltaics *Engineering* **5** 3358–80

[37] NREL 2010 NREL's black silicon increases solar cell efficiency by reducing reflected light *Report* NREL/FS-6A4-48772 NREL https://nrel.gov/docs/fy11osti/48772.pdf

[38] Oh J, Branz H M and Yuan H-C 2022 Efficient black silicon photovoltaic device with enhanced blue response *US Patent.* No. 11 251 318 B2

[39] Yuan H-C *et al* 2009 Efficient black silicon solar cell with a density-graded nanoporous surface: optical properties, performance limitations, and design rules *Appl. Phys. Lett.* **95** 123501

[40] Branz H M *et al* 2009 Nanostructured black silicon and the optical reflectance of graded-density surfaces *Appl. Phys. Lett.* **94** 231121

[41] Sprafke A N and Wehrspohn R B 2012 Light trapping concepts for photon management in solar cells *Green* **2** 177–87

[42] Law A M, Jones L O and Walls J M 2023 The performance and durability of anti-reflection coatings for solar module cover glass—a review *Sol. Energy* **261** 85–95

[43] Jansen H *et al* 1996 A survey on the reactive ion etching of silicon in microtechnology *J. Micromech. Microeng.* **6** 14–24

[44] Savin H *et al* 2015 Black silicon solar cells with interdigitated back-contacts achieve 22.1% efficiency *Nature Nanotech.* **10** 624–8
[45] Otto M *et al* 2012 Conformal Al_2O_3 coatings on black silicon by thermal ALD for surface passivation *Energy Procedia* **27** 361–4
[46] Oh J, Yuan H-C and Branz H M 2012 An 18.2%-efficient black-silicon solar cell achieved through control of carrier recombination in nanostructures *Nat. Nanotechnol.* **7** 743–8
[47] Chai J Y-H, Wong B T and Juodkazis S 2020 Black-silicon-assisted photovoltaic cells for better conversion efficiencies: a review on recent research and development efforts *Mater. Today Energy* **18** 100539
[48] XQE sensors & camera components *SIONYX* www.sionyx.com/pages/sensors (Accessed: 1 May 2025)
[49] Deng Y L and Juang Y-J 2014 Black silicon SERS substrate: effect of surface morphology on SERS detection and application of single algal cell analysis *Biosens. Bioelectron.* **53** 37–42
[50] Gervinskas G *et al* 2013 Surface-enhanced Raman scattering sensing on black silicon *Ann. Phys.* **525** 907–14
[51] Xu Z 2014 Black silicon and its applications in photovoltaics and biosensing *PhD Thesis* (University of Illinois at Urbana Champagne)
[52] Padrez Y and Golubewa 2024 Black silicon surface-enhanced Raman spectroscopy biosensors: current advances and prospects *Biosensors* **14** 453
[53] Bércesa Z *et al* 2014 Neural cell response to nanostructured biosensor surfaces *Procedia Eng.* **87** 971–4
[54] Black silicon sensors for molecular contamination *Melbourne Centre for Nanofabrication* https://nanomelbourne.com/case_study/black-silicon-sensors-for-molecular-contamination/ (Accessed: 1 May 2025)
[55] de la Mora M B *et al* 2013 *State of the Art in Biosensors—General Aspects* (InTech) pp 141–61 https://intechopen.com/books/state-of-the-art-in-biosensors-general-aspects/porous-silicon-biosensors
[56] Sainiemi L *et al* 2007 Rapid fabrication of high aspect ratio silicon nanopillars for chemical analysis *Nanotechnology* **18** 505303
[57] Dey A 2018 Semiconductor metal oxide gas sensors: a review *Mater. Sci. Eng.* B **229** 206–17
[58] Zappa D *et al* 2018 Metal oxide-based heterostructures for gas sensors—a review *Anal. Chim. Acta* **1039** 1–23
[59] Zhu L and Zeng W 2017 Room-temperature gas sensing of ZnO-based gas sensor: a review *Sensors Actuators* A **267** 242–61
[60] Goel N *et al* 2023 Metal oxide semiconductors for gas sensing *Eng. Rep.* **5** e12604
[61] Ayvazyan G *et al* 2023 NO_2 gas sensor based on pristine black silicon formed by reactive ion etching *Phys. Status Solidi-Rapid Res. Lett.* **17** 2300058
[62] Nanostructured black silicon *AZoNano* 18 September 2014 https://azonano.com/article.aspx ? ArticleID = 3746 (Accessed: 3 May 2025)
[63] Nanopatterned Black Silicon Shows Antibacterial Power, https://www.asianscientist.com/2013/12/in-the-lab/black-silicon-shows-antibacterial-power-2013/ (Accessed: 17 June 2025)
[64] Ivanova E P *et al* 2013 Bactericidal activity of black silicon *Nat. Commun.* **4** 1–11
[65] Narasimhan V *et al* 2017 Black silicon as a multifunctional material for medical implants: first demonstrated use in *in-vivo* intraocular pressure sensing *IEEE Transducers (Kaohsiung, Taiwan, 18–22 June)* pp 387–90

[66] Butler J *et al* 2023 Review of antimicrobial nanocoatings in medicine and dentistry: mechanisms of action, biocompatibility performance, safety, and benefits compared to antibiotics *ACS Nano* **17** 7064–92
[67] How cicada wings kill bacteria *Ask Nature* https://asknature.org/strategy/how-cicada-wings-kill-bacteria/ (Accessed: 4 May 2025)
[68] Ivanova E P *et al* 2020 The multi-faceted mechano-bacterial mechanism of nanostructured surfaces *PNAS* **117** 12598–605
[69] Conners D 2025 Cicada wings inspire new ideas for antibacterial products *Earth Sky* https://earthsky.org/earth/cicada-wings-inspire-new-ideas-for-antibacterial-products/ (Accessed: 4 May 2025)
[70] Bird D and Ravindra N M 2020 Transdermal drug delivery and patches—an overview *Med. Devices Sens.* **3** e10069
[71] Henry S *et al* 1998 Micromachined needles for the transdermal delivery of drugs *The Eleventh Ann. Int. Workshop Proc. on MEMS* pp 494–8
[72] Teo A L *et al* 2006 Transdermal microneedles for drug delivery applications *Mater. Sci. Eng.* B **132** 151–4
[73] Cheng W *et al* 2023 Fabrication of black silicon microneedle arrays for high drug loading *J. Funct. Biomater.* **14** 245
[74] Pavlova S G *et al* 2008 Multi-crystalline silicon as active medium for terahertz intracenter lasers *Physica* B **403** 535–8
[75] Hoyer P *et al* 2008 Terahertz emission from black silicon *Appl. Phys. Lett.* **93** 091106
[76] Qi D *et al* 2009 Simple approach to wafer-scale self-cleaning antireflective silicon surfaces *Langmuir* **25** 7769–72
[77] Nishiijima Y, Nishiijima H and Juodkazis S 2020 Black silicon as a highly efficient photo-thermal converter for snow/ice melting in early spring agriculture *Sol. Energy Mater. Sol. Cells* **217** 110706
[78] Nishiijima Y, Nishiijima H, Katkus T and Juodkazis S 2025 Efficient water evaporation using black silicon *ACS Sustain. Resour. Manage.* **2** 316–21
[79] Chai J Y H, Wong B T and Sunarso J 2023 An opto-electro-thermal model for black-silicon assisted photovoltaic cells in thermophotovoltaic applications *Photonics* **10** 565
[80] Azri A A K *et al* 2024 Development of surface-texturized black silicon through metal-assisted chemical etching and its application in the thermophotovoltaic field: a review and recommendation *Semicond. Sci. Technol.* **40** 013001
[81] Vlasov A A *et al* 2022 Machine vision system utilizing black silicon CMOS camera for through-silicon alignment *IEEE Trans Compon. Packag. Manuf. Technol.* **12** 2019–26
[82] Jansen H *et al* 1995 The black silicon method IV: the fabrication of three-dimensional structures in silicon with high aspect ratios for scanning probe microscopy and other applications *Proc. IEEE Micro Electro Mechanical Systems (Amsterdam, 29 January–2 February)*
[83] Shiri R *et al* 2019 Highly absorptive pupil mask fabricated with black-silicon *Proc. SPIE* **11116** 111161H

www.ingramcontent.com/pod-product-compliance
Ingram Content Group UK Ltd.
Pitfield, Milton Keynes, MK11 3LW, UK
UKHW051853100926
8556IPUK00005B/599

9 780750 332453